T0091535

KIAS Springer Series in Mathematics

Volume 1

The KIAS Springer Series in Mathematics publishes original content in the form of high level research monographs, lecture notes, proceedings and contributed volumes as well as advanced textbooks in English language only, in any field of Pure and Applied Mathematics. The books in the Series are connected to the research activities carried out by the Korea Institute for Advanced Study (KIAS), and will discuss recent results and analyze new trends in mathematics and its applications. The Series is aimed at providing useful reference material to academics and researchers at an international level.

Nam-Gyu Kang · Jaigyoung Choe ·
Kyeongsu Choi · Sang-hyun Kim
Editors

Recent Progress
in Mathematics

 Springer

Editors
Nam-Gyu Kang
School of Mathematics
Korea Institute for Advanced Study
Seoul, Korea (Republic of)

Kyeongsu Choi
School of Mathematics
Korea Institute for Advanced Study
Seoul, Korea (Republic of)

Jaigyoung Choe
Korea Institute for Advanced Study
Seoul, Korea (Republic of)

Sang-hyun Kim
School of Mathematics
Korea Institute for Advanced Study
Seoul, Korea (Republic of)

ISSN 2731-5142   ISSN 2731-5150 (electronic)
KIAS Springer Series in Mathematics
ISBN 978-981-19-3710-1   ISBN 978-981-19-3708-8 (eBook)
https://doi.org/10.1007/978-981-19-3708-8

Mathematics Subject Classification: 14N35, 14C17, 35Q30, 35Q35, 35J05, 35P05, 53C44, 58K55, 37F10

This Springer imprint is published by the registered company Springer Nature Singapore Pte Ltd.
The registered company address is: 152 Beach Road, #21-01/04 Gateway East, Singapore 189721,
Singapore

Preface

The Korea Institute for Advanced Study (KIAS) organized the Quadranscentennial KIAS Lectures for her 25th anniversary and KIAS Expositions Lectures in 2021. KIAS Expositions contains the expository articles that formed these lectures. The first volume of KIAS Expositions contains four expository articles and one research paper in essential areas of mathematics such as algebraic geometry, topology, partial differential equations, Riemannian geometry, and harmonic analysis. These articles present not only the authors' current research but also explore the history of the current problems and conjecture their future directions.

The KIAS Expositions was launched early in 2021 as an official journal of the KIAS. It was designed for high-quality expository articles in all areas of mathematics. Soon after its launch, KIAS and Springer agreed to turn the KIAS Expositions into the KIAS Springer Series in Mathematics to expand the journal into a book series. The Series will publish original contents in the form of high-level research monographs, lecture notes, proceedings, contributed volumes, and advanced textbooks in any field of Pure and Applied Mathematics. Thus, the first volume of KIAS Expositions became the first volume of the KIAS Springer Series. This volume includes the expository articles by Young-Hoon Kiem, Dongho Chae, Simon Brendle, Hyeonbae Kang, and the research paper by Danny Calegari.

In his expository paper "Enumerative geometry, before and after string theory," Young-Hoon Kiem offers a glimpse into this story about enumerative geometry and string theory. In around 1990, a group of string theorists applied the mirror symmetry in string theory to enumerative geometry problems in algebraic geometry. For three decades since then, one of the most fruitful interactions between mathematicians and physicists has happened in enumerative geometry. Kiem discusses classical enumerative geometry before string theory and improvements after string theory, as well as some recent advances in quantum singularity theory: Donaldson–Thomas theory for Calabi–Yau 4-folds and Vafa–Witten invariant.

In this expository article "On the singularity problem for the Euler equations" by Dongho Chae, the author discusses the finite-time singularity problem for the three-dimensional incompressible Euler equation. Solutions with initial data in $H^k(\mathbb{R}^3)$ with $k > 5/2$ are well-posed local in time. However, it is a wide-open problem if a smooth solution develops singularities. The author reviews the blow-up criterion by Beale–Kato–Majda, and results related to the Type I blow-up. He also addresses the two-dimensional Boussinesq equation, which is a good model problem for the axisymmetric three-dimensional incompressible Euler equation.

In the survey paper "Singularity models in the three-dimensional Ricci flow" by Simon Brendle, the author addresses a complete classification of all the singularity models of the Ricci flow on three-dimensional manifolds. The author discusses the recent development in the study of noncollapsed ancient solutions, with an emphasis on steady gradient Ricci solitons. By using recently established theorems, the author provides an alternative proof for the uniqueness of noncollapsed steady gradient Ricci solitons in dimension three, which was originally proved by the author in 2012.

In his expository article "Spectral geometry and analysis of the Neumann–Poincaré operator, a review," Hyeonbae Kang reviews some of the recent development in the spectral theory of Neumann–Poincaré (NP) operator. As an integral operator on the boundary of a domain in Euclidean space, the NP operator was introduced to solve Dirichlet or Neumann boundary value problems. Hyeonbae Kang discusses the visibility and invisibility of the NP operator via polarization tensors, the decay rate of eigenvalues and the surface localization of plasmon, the singular geometry and spectrum on polygonal domains, the analysis of stress using the spectral theory, and the structure of elastic NP operator.

In his research paper "Sausages and butcher paper," Danny Calegari presents two novel ways to understand the structure of the shift locus, which is defined as the space of normalized complex polynomials with a certain degree whose restrictions on the Julia sets are conjugate to shifts on finite alphabets. Those are combinatorial on one side and algebro-geometric on the other side. As a consequence, the author can realize the shift locus as a complex of spaces over a contractible $\tilde{A}_{d-2}$ building.

Seoul, Korea (Republic of)   Jaigyoung Choe
May 2022   Kyeongsu Choi
                                                                               Nam-Gyu Kang
                                                                               Sang-hyun Kim

Contents

Enumerative Geometry, Before and After String Theory

Young-Hoon Kiem

Abstract Throughout its long history, algebraic geometry has been enriched by the influences of invariant theory, abstract algebra, number theory, topology, complex analysis, symplectic geometry, category theory and so on. One of the most recent influences came from string theory circa 1990 when a group of string theorists applied the mirror symmetry in string theory to enumerative geometry problems in algebraic geometry. During the past three decades, we have witnessed one of the most fruitful interactions between mathematicians and physicists in enumerative geometry. In this informal expository article, I would like to offer a glimpse into this story about enumerative geometry and string theory.

Keywords Enumerative geometry

1 Introduction

According to Hermann Schubert in 1874, enumerative geometry is the study about questions like

Question 1 How many geometric figures of fixed type satisfy certain given conditions? ☐

It is one of the oldest branches of mathematics and dates back to at least 200 BCE when Apollonius raised the question:

Question 2 How many circles in plane are tangent to three given circles? ☐

To simplify, let us assume that the disks bounded by the three circles are disjoint from one another. If a circle X is tangent to the given three circles A_1, A_2, A_3, then each

---

This article was written in June 2020.

Y.-H. Kiem (✉)
Department of Mathematics and Research Institute of Mathematics, Seoul National University, Seoul 08826, Korea
e-mail: kiem@snu.ac.kr

© The Author(s), under exclusive license to Springer Nature Singapore Pte Ltd. 2022   1
N.-G. Kang et al. (eds.), *Recent Progress in Mathematics*, KIAS Springer Series
in Mathematics 1, https://doi.org/10.1007/978-981-19-3708-8_1

of the three circles may lie inside or outside of X, and hence there are 8 possibilities. Obviously X is determined by three numbers, namely the coordinates (a, b) of the center and the radius r. The tangency conditions for the three circles give us three quadratic equations,

$$(a - a_i)^2 + (b - b_i)^2 = (r \pm r_i)^2, \quad i = 1, 2, 3 \tag{1}$$

where (a_i, b_i) is the center and r_i is the radius of A_i. The sign in the right hand side is determined by the choice among eight possibilities of lying inside or outside. It is easy to solve (1) and see that the answer to Apollonius problem is 8. Of course, one may solve this problem by plane geometry without the help of algebra, starting with the fact that the locus[1] of centers of circles tangent to two given circles is a hyperbola and Apollonius asks you to look for the intersection of three hyperbolas.

A circle is the zero locus in plane of a quadratic polynomial of the form $x^2 + y^2 + \cdots$ where $\cdots$ means lower order terms. More generally, the zero locus in plane of any quadratic polynomial is called a *conic* (curve) because they are obtained by intersecting a circular cone with planes.[2] A moment's thought may convince the reader that it requires 5 numbers to determine a conic. For instance, to determine an ellipse, we need to know its center, the lengths of long and short axes and the slope of the long (or short) axis. So we need five equations to determine these numbers (because we may delete one variable by using one equation) and an equation for the defining parameters usually comes from a geometric constraint like tangency. In 1848, J. Steiner thought about a seemingly naive generalization of Apollonius' question.

Question 3 How many conics in plane are tangent to five given conics?   □

Steiner correctly argued that the set of conics tangent to a given conic is the zero locus of degree 6 (sextic) polynomial in 5 variables[3] and thus we are looking for the number of solutions of a system of 5 sextic equations in 5 variables. So Steiner's problem is related to the algebraic question:

Question 4 Given a system of 5 general[4] sextic polynomial equations in 5 variables, how many solutions are there? More generally, given a system of r polynomial

---

[1] The meaning of *locus* is the same as *set*, often with a touch of geometric flavor (sometimes subvarieties, subschemes or substacks). The *zero locus* of polynomials $f_1, \cdots, f_r$ means the set of points p whose coordinates satisfy all the polynomial equations $f_1(p) = 0, \cdots, f_r(p) = 0$.

[2] A more appropriate term for a *conic* from the perspective of modern algebraic geometry is a *quadratic curve*.

[3] It suffices to consider a pencil $\{X_t\}_{t \in \mathbb{P}^1}$ of conics and find the number of conics in this pencil which are tangent to a given smooth conic C. Here a curve is smooth if and only if the tangent space at each point is 1-dimensional. In particular, a plane curve C defined by a polynomial f is smooth if and only if the gradient vector of f at every point in C is nonzero. The intersection $X_t \cap C$ is a divisor of degree 4 in $C \cong \mathbb{P}^1$ and hence gives us a a degree 4 map $C \to \mathbb{P}^1$. There are precisely 6 ramification points by the Hurwitz formula $(6 = 2 \cdot 4 - 2)$. Hence the pencil intersects with the locus of conics tangent to C at 6 points.

[4] We want to avoid extreme cases which behave badly, like the case where there are infinitely many solutions. The word *general* means that we are excluding such unfortunate possibilities.

equations of degrees $d_1, \cdots, d_r$ respectively, in r variables, how many solutions are there?   □

For an insight, it is always a good idea to consider a simplified version of a difficult problem.

Question 5 If you have 2 general polynomial equations of degrees m, n in 2 variables, how many common solutions do they have?   □

Geometrically speaking, two general curves of degrees m, n in plane meet at mn points.[5] Why? Since the coordinates of intersection points are the solutions of polynomial equations which vary continuously as we vary the coefficients of the polynomials, the number of intersection points remains constant as we vary the coefficients continuously and hence we can deform the equations continuously until we reach

$$\prod_{0 \le i < m} (x - i), \quad \prod_{0 \le j < n} (y - j) \tag{2}$$

without changing the number of intersection points. Obviously the number of intersection points of the curves defined by (2) is mn. This method of deforming or degenerating into simple cases is one of the most important techniques in mathematics and in fact, a huge part of geometry and topology is based on this single technique of stability under degeneration or deformation.[6]

The above geometric argument looks simple but actually requires many pages of justification to make it precise. Can you prove it algebraically? The starting point is the simple observation that a polynomial factorizes into a product of irreducible polynomials.[7] If two polynomials

$$f(x) = a_0 x^m + a_1 x^{m-1} + \cdots + a_{m-1} x + a_m, \tag{3}$$

$$g(x) = b_0 x^n + b_1 x^{n-1} + \cdots + b_{n-1} x + b_n$$

in one variable of degrees m, n share a root, then they have a common factor and hence their least common multiple has degree less than mn. So we can find polynomials h, k with $\deg h < n$, $\deg k < m$ such that $hf + kg = 0$. If we think of the coefficients of h, k as variables and write the equation $hf + kg = 0$ as a system of linear equations in the coefficients, we obtain an $(m + n) \times (m + n)$ matrix

---

Mathematically, it means that we are choosing the parameters (coefficients of the equations in this case) in an open dense subset (in the Zariski topology).

[5] The degree of a plane curve is the degree of the defining polynomial.

[6] The idea is the same but their names may vary in different branches of mathematics.

[7] For a field F, a nonzero polynomial with coefficients in F is the product of irreducible polynomials in a unique way up to the order of multiplication and units.

$$\begin{pmatrix} a_0 & a_1 & a_2 & \cdots & a_m & 0 & 0 & 0 \\ 0 & a_0 & a_1 & \cdots & a_{m-1} & a_m & 0 & 0 \\ \cdots & \cdots & \cdots & \cdots & \cdots & \cdots & \cdots & \cdots \\ 0 & 0 & \cdots & 0 & a_0 & \cdots & a_{m-1} & a_m \\ b_0 & b_1 & b_2 & \cdots & b_n & 0 & 0 & 0 \\ 0 & b_0 & b_1 & \cdots & b_{n-1} & b_n & 0 & 0 \\ \cdots & \cdots & \cdots & \cdots & \cdots & \cdots & \cdots & \cdots \\ 0 & 0 & \cdots & 0 & b_0 & \cdots & b_{n-1} & b_n \end{pmatrix} \tag{4}$$

whose determinant is called the *resultant* of f, g and denoted by $R(f, g)$. The two polynomials f, g have a common root if and only if $hf + kg = 0$ for some nonzero polynomials h, k with $\deg h < n$, $\deg k < m$, which is equivalent to saying that their resultant $R(f, g)$ is zero.[8]

Now if $f(x, y)$ and $g(x, y)$ are polynomials in two variables, we may write them as (3) with a_i, b_j being polynomials in y of degrees $\leq i$ and $\leq j$ respectively. It is not hard to see that the resultant $R(f, g)$ is a polynomial in y of degree mn for general f, g. In the general case where two curves defined by f and g intersect only at finitely many points, we may pick the coordinate system so that the y coordinates of the intersection points are all distinct, which means that the system $f(x, y) = 0 = g(x, y)$ has at most one solution if y is fixed. For the real numbers, the only conclusion we can draw is that there are at most mn solutions to the system $f(x, y) = 0 = g(x, y)$ because $R(f, g) = 0$ has at most mn solutions.

We can do much better with complex numbers. As there is a solution for any polynomial equation in one variable, the system $f(x, y) = 0 = g(x, y)$ has exactly one solution of the form (x, y_0), for each (simple) root y_0 of $R(f, g) = 0$. Thus if $R(f, g) = 0$ has k solutions, then so does the system $f(x, y) = 0 = g(x, y)$. More-over, if we use complex numbers, for general f, g, we can see that $R(f, g) = 0$ has exactly mn distinct solutions and hence the system $f(x, y) = 0 = g(x, y)$ has exactly mn solutions. Therefore the answer to Question 5 is mn over $\mathbb{C}$.

Remark As you can see, for a consistent theory, it is much better to use the complex number field $\mathbb{C}$ (or an algebraically closed field) instead of real numbers. So from now on, we will always use *complex numbers*[9] as our scalars.   □

Motivated by our answer to Question 5, you may expect that general polyno-mial equations $f_1 = \cdots = f_r = 0$ of degrees $d_1, \cdots, d_r$ in r variables should have $\prod_{i=1}^{r} d_i$ solutions over $\mathbb{C}$. Indeed this is true although it requires a more sophisticated

---

[8] In fact, $R(f, g)$ is a constant multiple of $\prod_{i,j}(\alpha_i - \beta_j)$ where $\{\alpha_i\}_{1 \leq i \leq m}$ (resp. $\{\beta_j\}_{1 \leq j \leq n}$) are the roots of $f = 0$ (resp. $g = 0$).

[9] For real numbers, one can first look for complex solutions and see which of them are invariant under conjugation.

proof.[10] In particular, 5 general sextic polynomial equations in 5 variables should have $6^5 = 7776$ solutions and this number is Steiner's answer to Question 3.

It turned out that 7776 is the correct answer to Question 4 but not to Question 3. Why? The problem is that the 5 sextic polynomials for the tangency of 5 given conics are *never* general. The correct answer to Question 3 was found to be 3264 by Chasles in 1864 but a mathematically rigorous proof was provided by Fulton and MacPherson, only in 1978! It took more than a century to settle such a naive looking problem.

Why 3264 instead of 7776? In the space of all conics, namely in the space of all quadratic polynomials in two variables up to constant, there is a subspace of double lines, namely the set of squares of linear polynomials. Double lines are tangent to any conic although we are not interested in them. So in the space of conics, the subset Z of conics tangent to the 5 given conics consists of 3264 honest smooth conics and also infinitely many double lines. If we deform the defining equations of the 5 sextic polynomials defining the subset Z, extra 4512 intersection points arise from the locus of double lines which we do not want to consider.[11]

---

[10] By Hilbert's Nullstellensatz, the number of solutions is the same as the dimension

$$\dim \mathbb{C}[x_1, \cdots, x_r]/\langle f_1, \cdots, f_r\rangle.$$

To compute this dimension, by adding an extra variable x_0, homogenize $f_1, \cdots, f_r$ into homogeneous polynomials $F_1, \cdots, F_r$ of degrees $d_1, \cdots, d_r$. For ℓ large enough, we have an isomorphism

$$\mathbb{C}[x_1, \cdots, x_r]/\langle f_1, \cdots, f_r\rangle \cong \mathbb{C}[x_0, \cdots, x_r]_\ell/\langle F_1, \cdots, F_r\rangle_\ell$$

where the subscript ℓ stands for the degree ℓ homogeneous part. The dimension of the right hand side can be computed by the Koszul complex

$$0 \longleftarrow \mathbb{C}[x_0, \cdots, x_r]_\ell/\langle F_1, \cdots, F_r\rangle_\ell \longleftarrow \mathbb{C}[x_0, \cdots, x_r]_\ell \overset{(F_1, \cdots, F_r)}{\longleftarrow} \oplus_{1 \le i \le r} \mathbb{C}[x_0, \cdots, x_r]_{\ell - d_i}$$

$$\longleftarrow \oplus_{1 \le i < j \le r} \mathbb{C}[x_0, \cdots, x_r]_{\ell - d_i - d_j} \longleftarrow \cdots \longleftarrow \mathbb{C}[x_0, \cdots, x_r]_{\ell - d_1 - \cdots - d_r} \longleftarrow 0$$

together with the identity

$$\prod_{i=1}^{r} d_i = \sum_{i=0}^{r} (-1)^i \sum_{1 \le j_1 < \cdots < j_i \le r} \binom{r + \ell - d_{j_1} - \cdots - d_{j_i}}{r}.$$

[11] To remove the contribution from the locus of double lines, we blow up the space of conics along the locus of double lines (Veronese surface) and then compute the intersection number $(6H - 2E)^5 = 3264$ where H is the hyperplane class and E is the exceptional divisor. The space of conics $\mathbb{P}^5$ is the moduli space of ideal sheaves (Hilbert scheme) and hence Steiner's answer 7776 is a Donaldson-Thomas (DT) invariant in algebraic geometry or BPS state counting in physics terms. The blowup of $\mathbb{P}^5$ along the Veronese surface is the moduli space of stable maps to $\mathbb{P}^2$ of degree 2 and hence Chasles's answer 3264 is the Gromov-Witten (GW) invariant in modern terms. The GW/DT correspondence, or the MNOP conjecture, claims that the GW invariants have the same amount of information on enumerative geometry of curves as the DT invariants. It is remarkable that the first nontrivial case of the GW/DT correspondence or the gauge/string duality already appeared in mathematics, 150 years before string theory.

Enumerative geometry is about natural questions on curves and surfaces etc but solving such problems often requires an immense amount of technical training and ingenuities.

In his book *What is mathematics?*, Richard Courant wrote:

Understanding of mathematics cannot be transmitted by painless entertainment any more than education in music can be brought by the most brilliant journalism to those who never have listened intensively. Actual contact with the content of living mathematics is necessary.

I agree that the best way to appreciate mathematics is walking through carefully selected problems. In this note, I will try to convey the ideas in enumerative geometry through explicit examples as much as possible. If you enjoyed so far, you will find more interesting stories below.

In the subsequent sections, we will see more of classical enumerative geometry problems, its connection with string theory and recent developments. Everything in this note was taken from books, papers and memories. No part of this note is original.

2   Classical Enumerative Geometry

It is great fun to play with enumerative geometry problems.

2.1   *Schubert Calculus*

Schubert calculus is the enumerative geometry of lines, planes and more generally geometric figures defined by linear equations and sometimes quadratic polynomial equations. For instance, a single linear equation in 3 variables defines a plane while two linear equations usually define a line. Let us consider the following question which you may find in a standard undergraduate algebraic geometry textbook.

Question 6 How many lines in 3 dimensional space meet 4 given general lines? □

A line in space $\{(x, y, z) \mid x, y, z \text{ scalars}\}$ is the intersection of two planes

$$ax + by + cz + d = 0, \quad ex + fy + gz + h = 0,$$

$$(a, b, c) \nparallel (e, f, g).$$

Since the equations are determined by the coefficients, we may just record them in the matrix form

$$A = \begin{pmatrix} a & b & c & d \\ e & f & g & h \end{pmatrix} \in M_2^{2 \times 4} \tag{5}$$

where $M^{2 \times 4}$ denotes the set of 2×4 matrices and the subscript 2 denotes the open subset of matrices of rank 2. It is easy to see that two such 2×4 matrices A and A' represent the same line if and only if $A' = gA$ for an invertible 2×2 matrix $g \in GL_2$.

We denote by GL_n the set of invertible $n \times n$ matrices. The identity matrix I lies in GL_n and if $\xi, \xi' \in GL_n$, then $\xi^{-1}, \xi\xi' \in GL_n$, i.e. GL_n is a *group*. The closed subset of $n \times n$ matrices ξ with determinant 1 is denoted by SL_n which is a subgroup of GL_n because the determinant is multiplicative. Constant multiples of I form another subgroup Z of GL_n which is complementary to SL_n.[12]

For $A \in M_2^{2 \times 4}$, the set $\{\xi A \mid \xi \in GL_2\}$ is called the *orbit* of A and the set of orbits in $M_2^{2 \times 4}$ is denoted by

$$GL_2 \backslash M_2^{2 \times 4} =: Gr(2, 4).$$

Therefore the set X of lines in space is an open subset of $Gr(2, 4)$, namely the set of orbits of matrices whose left 2×3 submatrix has rank 2. What is the geometry of X and $Gr(2, 4)$, the orbit spaces by the action of GL_2? Let us first consider the orbits by SL_2 and later by Z. By the multiplicative property $\det(\xi\xi') = \det(\xi)\det(\xi')$, we have a map

$$\Phi : SL_2 \backslash M_2^{2 \times 4} \longrightarrow \mathbb{C}^6, \quad A \mapsto (\Phi_{12}, \Phi_{13}, \Phi_{14}, \Phi_{23}, \Phi_{24}, \Phi_{34})$$

whose coordinates are the 2×2 minor determinants[13]

$$\Phi_{12}(A) = af - be, \quad \Phi_{13}(A) = ag - ce, \quad \Phi_{14}(A) = ah - de, \tag{6}$$

$$\Phi_{23}(A) = bg - cf, \quad \Phi_{24}(A) = bh - df, \quad \Phi_{34}(A) = ch - gd$$

---

[12] To be precise, $GL_n = SL_n \times_{\mu_n} Z = SL_n \times Z/\mu_n$, where $\mu_n = \{\zeta \mid \zeta^n = 1\}$.

[13] An affine variety (a subset of a vector space defined as the zero locus of a finitely many polynomials) is entirely determined by its coordinate ring of polynomial functions (functions given by polynomials) up to isomorphism. The coordinate ring for the quotient $SL_2 \backslash M^{2 \times 4}$ is the ring of polynomial functions on the vector space $M^{2 \times 4}$ which are invariant under the action of SL_2. In the 19th century, one of the most intensively studied branches of mathematics is Invariant Theory whose goal is to find the ring of polynomials invariant under a group action. Many branches in current mathematics arose from invariant theory, such as linear algebra, algebraic group theory, representation theory, geometric invariant theory and more. Also the influence of invariant theory on quantum physics is pervasive and evident. The fundamental theorems in invariant theory for SL_n tell us that the ring of SL_2-invariant polynomial functions on $M^{2 \times 4}$ is generated by the 2×2 minor determinants (6) with the Plucker relation (7).

of the 2×4 matrix (5). The six polynomials satisfy only one algebraic relation

$$\Phi_{12}\Phi_{34} - \Phi_{13}\Phi_{24} + \Phi_{14}\Phi_{23} = 0. \tag{7}$$

It is easy to see that the map Φ is injective and hence the set of SL_2 orbits in $M_2^{2\times 4}$ is a quadric hypersurface

$$Q = \{(z_1, \cdots, z_6) \,|\, z_1 z_6 - z_2 z_5 + z_3 z_4 = 0, \text{ not all } z_i \text{ are } 0\} \subset \mathbb{C}^6 - \{0\}$$

and the set X of lines in $\mathbb{C}^3$ is identified with the open subset of the quadric hypersurface Q, defined by $(z_1, z_2, z_4) \neq (0, 0, 0)$. Here a *hypersurface* means the zero set of a single polynomial. A *hyperplane* will mean a hypersurface defined by a linear polynomial.

Let us now consider the action of the center $Z = \{tI \,|\, t \in \mathbb{C}^*\}$. We see that $\Phi(tA) = t^2\Phi(A)$ and hence we have to consider the orbit space

$$\mathbb{C}^*\backslash(\mathbb{C}^6 - \{0\}) =: \mathbb{P}^5$$

where the constants $t \in \mathbb{C}^*$ act on $\mathbb{C}^6 - \{0\}$ by scalar multiplication. This orbit space is called the 5 dimensional *projective space* and let us denote by $[z_1, \cdots, z_6]$ the orbit of $(z_1, \cdots, z_6) \in \mathbb{C}^6 - \{0\}$.[14] We call $z_1, \cdots, z_6$ *homogeneous coordinates* of $\mathbb{P}^5$.

Combining the discussions above, we find that the space X of all lines in space is the open subset in the quadric hypersurface

$$Gr(2, 4) = \{[z_1, \cdots, z_6] \in \mathbb{P}^5 \,|\, z_1 z_6 - z_2 z_5 + z_3 z_4 = 0\} \subset \mathbb{P}^5,$$

---

[14] The projective space $\mathbb{P}^5$ is a smooth manifold of (complex) dimension 5 with the quotient topology by the map
$$\mathbb{C}^6 - \{0\} \to \mathbb{C}^*\backslash(\mathbb{C}^6 - \{0\}) = \mathbb{P}^5, \quad (z_1, \cdots, z_6) \mapsto [z_1, \cdots, z_6].$$
$\mathbb{P}^5$ is compact as there is a surjective continuous map $S^5 \subset \mathbb{C}^6 - \{0\} \to \mathbb{P}^5$ from the unit sphere in $\mathbb{C}^6$ which is a closed and bounded subset. Recall that given a surjective map $f : X \to Y$ from a topological space X to a set Y, the quotient topology of Y is obtained by declaring that a subset U of Y is open if and only if $f^{-1}(U)$ is open. All orbits of points $(z_1, \cdots, z_6)$ with $z_1 \neq 0$ in $\mathbb{C}^6 - \{0\}$ meet the hyperplane $z_1 = 1$ at a unique point and we may identify this open subset $U_1 = (z_1 \neq 0)$ of $\mathbb{P}^5$ with $\mathbb{C}^5 = \{(1, z_2, \cdots, z_6) \,|\, z_i \in \mathbb{C}\}$. Ignoring z_1, we find that its complement $(z_1 = 0)$ is
$$\mathbb{P}^5 - U_1 = \mathbb{C}^*\backslash(\mathbb{C}^5 - \{0\}) =: \mathbb{P}^4.$$
By induction, we thus find that
$$\mathbb{P}^5 = \mathbb{C}^5 \sqcup \mathbb{P}^4 = \mathbb{C}^5 \sqcup \mathbb{C}^4 \sqcup \mathbb{P}^3 = \cdots = \mathbb{C}^5 \sqcup \mathbb{C}^4 \sqcup \cdots \sqcup \mathbb{C} \sqcup pt.$$
Hence the projective space $\mathbb{P}^5$ is a compactification of $\mathbb{C}^5$, meaning that it is a compact space containing $\mathbb{C}^5$ as a dense open subset. The open subsets $U_i = (z_i \neq 0)$ are charts of $\mathbb{P}^5$ which make $\mathbb{P}^5$ a smooth manifold.

defined by $(z_1, z_2, z_4) \neq (0, 0, 0)$.[15] It is a 4 dimensional space as you can expect.[16] This wraps up our description of the space X of all lines in space.

Given a line

$$\frac{x - p}{\ell} = \frac{y - q}{m} = \frac{z - r}{n} \tag{8}$$

in space, a line determined by (5) meets (8) if and only if

$$\det A \begin{pmatrix} \ell & p \\ m & q \\ n & r \\ 0 & 1 \end{pmatrix} = 0. \tag{9}$$

It is easy to see that (9) equals

$$(\ell q - pm)\Phi_{12} + (\ell r - pn)\Phi_{13} + \ell\Phi_{14} + (mr - qn)\Phi_{23} + m\Phi_{24} + n\Phi_{34} = 0,$$

a linear combination of $\{\Phi_{ij}\}$. So we see that the set of lines meeting (8) is a hyperplane in $\mathbb{P}^5$ defined by

$$(\ell q - pm)z_1 + (\ell r - pn)z_2 + \ell z_3 + (mr - qn)z_4 + m z_5 + n z_6 = 0. \tag{10}$$

For instance, consider the 4 lines given by

$$\begin{pmatrix} \ell & p \\ m & q \\ n & r \\ 0 & 1 \end{pmatrix} = \begin{pmatrix} 0 & 0 \\ 0 & 0 \\ 1 & 0 \\ 0 & 1 \end{pmatrix}, \begin{pmatrix} 1 & 0 \\ 0 & 0 \\ 0 & 0 \\ 0 & 1 \end{pmatrix}, \begin{pmatrix} 0 & 0 \\ 1 & 0 \\ -1 & 1 \\ 0 & 1 \end{pmatrix}, \begin{pmatrix} 1 & 1 \\ -1 & 0 \\ 0 & 0 \\ 0 & 1 \end{pmatrix}. \tag{11}$$

Then (10) gives us

$$z_6 = z_3 = z_4 + z_5 - z_6 = z_1 + z_3 - z_5 = 0.$$

Together with the quadratic equation $z_1 z_6 - z_2 z_5 + z_3 z_4 = 0$, we find that there are exactly two solutions

$$[z_1, \cdots, z_6] = [1, 0, 0, -1, 1, 0], [0, 1, 0, 0, 0, 0] \in X \subset Gr(2, 4) \subset \mathbb{P}^5.$$

So the set of lines meeting all 4 given lines is the intersection of the 4 hyperplanes (of lines meeting each of the given 4 lines) with X in $\mathbb{P}^5$. For a general choice of 4 hyperplanes $H_1, \cdots, H_4$ of $\mathbb{P}^5$, the intersection

---

[15] The set of lines in a 3-dimensional vector space is thus a rank 2 vector bundle over $\mathbb{P}^2$.

[16] To specify a line L, we only need 2 parameters for the direction (namely a point in $\mathbb{P}^2$) of L and 2 parameters for the intersection point of L with the plane orthogonal to L and passing through the origin (rank 2 bundle).

$$Gr(2, 4) \cap H_1 \cap \cdots \cap H_4 \tag{12}$$

is disjoint from the lower dimensional subvarieties $Gr(2, 4) - X$ and $(z_1 = 0) = \mathbb{P}^5 - \mathbb{C}^5$. Hence finding the set (12) amounts to solving a system of 4 linear equations and 1 quadratic equation in 5 variables. By eliminating one variable by using one equation at a time (or by Question 4), we find that (12) consists of precisely 2 points and equals $X \cap H_1 \cap \cdots \cap H_4$.

As we've seen an example where there are exactly 2 lines meeting all 4 given lines,[17] we conclude that the answer to Question 6 is 2.

Is there an easier way to see it? Once again, try to convince yourself that if you vary your 4 given lines $L_1, \cdots, L_4$ continuously, the lines L meeting all of them will vary continuously. So we may consider a special configuration of lines. For instance, suppose $L_1 \cap L_2 = \{p_{12}\} \neq \emptyset$ and $L_3 \cap L_4 = \{p_{34}\} \neq \emptyset$. Let P_{12} (resp. P_{34}) denote the unique plane containing L_1 and L_2 (resp. L_3 and L_4). Then it is easy to see that the line $P_{12} \cap P_{34}$ and the line joining p_{12} and p_{34} are the only two lines meeting all 4 lines $\{L_i\}$. In fact, (11) is an example of such a configuration.

2.2 Enumerating Rational Curves in Plane

As we saw above, $\mathbb{P}^n$ is a compactification of the vector space $\mathbb{C}^n$ and a curve in $\mathbb{C}^n$ is compactified in $\mathbb{P}^n$ by taking the closure. If one is interested in finding curves in plane $\mathbb{C}^2$, it makes sense to find compact curves in $\mathbb{P}^2$. In this subsection, we enumerate curves in the projective plane $\mathbb{P}^2$ which are the images of polynomial maps $f : \mathbb{P}^1 \to \mathbb{P}^2$.

An *irreducible rational curve* in $\mathbb{P}^n$ of degree $d \geq 1$ is the image $f(\mathbb{P}^1)$ of a *polynomial map*[18]

$$f : \mathbb{P}^1 \to \mathbb{P}^n, \quad f([t, s]) = [f_0(t, s), \cdots, f_n(t, s)] \tag{13}$$

for homogeneous polynomials[19] $f_0, \cdots, f_n \in \mathbb{C}[t, s]_d$ of degree d (without a common factor), which is generically injective, i.e. there exists a finite set $A \subsetneq f(\mathbb{P}^1)$ such that f is injective on $\mathbb{P}^1 - f^{-1}(A)$.

A curve in $\mathbb{P}^2$ of degree 1 is a line $\mathbb{P}^1$ and hence an irreducible rational curve. A conic is defined by a homogeneous quadratic polynomial in z_0, z_1, z_2 and hence

---

[17] The Grassmannian is irreducible and the condition that (12) avoid $Gr(2, 4) - X$ and $(z_1 = 0)$ is open.

[18] A polynomial map is more often called a regular map or a morphism in algebraic geometry.

[19] A homogeneous polynomial is a polynomial whose nonzero monomials have the same degree. A polynomial $f \in \mathbb{C}[x_1, \cdots, x_r]$ is homogeneous of degree d if and only if $f(tx_1, \cdots, tx_r) = t^d f(x_1, \cdots, x_r)$ for $t \in \mathbb{C}^*$. The vector space of homogeneous polynomials of degree d in the polynomial ring $\mathbb{C}[x_1, \cdots, x_r]$ together with 0 is denoted by $\mathbb{C}[x_1, \cdots, x_r]_d$.

a symmetric 3×3 matrix. By linear algebra, we can find a coordinate system such that the conic is isomorphic to

1. $z_0^2 = 0$ (rank 1, double line), or
2. $z_0 z_1 = 0$ (rank 2, two lines), or
3. $z_0 z_2 = z_1^2$ (rank 3, smooth).

The last case is the image of the polynomial map

$$\mathbb{P}^1 \longrightarrow \mathbb{P}^2, \quad [t, s] \mapsto [t^2, ts, s^2].$$

Therefore any smooth curve in $\mathbb{P}^2$ of degree ≤ 2 is irreducible rational.

By definition, a polynomial map $f : \mathbb{P}^1 \to \mathbb{P}^n$ of degree d is determined by a choice of $n + 1$ homogeneous polynomials $f_0, \cdots, f_n \in \mathbb{C}[t, s]_d$. Writing

$$f_j(t, s) = \sum_{0 \leq i \leq d} a_{ij} t^{d-i} s^i, \tag{14}$$

we obtain a $(d + 1) \times (n + 1)$ matrix

$$A = (a_{ij}) \in M^{(d+1) \times (n+1)} = \mathbb{C}^{(d+1)(n+1)}$$

whose $\mathbb{C}^*$-orbit $[A] \in \mathbb{P}^{(d+1)(n+1)-1}$ determines f. The image $f(\mathbb{P}^1)$ does not change even if f is composed with a fractional linear transformation $\eta : \mathbb{P}^1 \to \mathbb{P}^1$.[20] Hence the set $R_d(\mathbb{P}^n)$ of all irreducible rational curves in $\mathbb{P}^n$ of degree d is an open subset in the orbit space

$$SL_2 \backslash \mathbb{P} M^{(d+1) \times (n+1)} = SL_2 \backslash \mathbb{P}^{(d+1)(n+1)-1} \tag{15}$$

whose dimension is $(d + 1)(n + 1) - 4$. In particular, when $n = 2$, the dimension of the space of all irreducible rational plane curves of degree d is $3d - 1$.

Given a collection of objects and an equivalence relation, their *moduli space* is the set of equivalence classes of objects in the collection. For example, the moduli space of circles in $\mathbb{R}^2$ is $\mathbb{R}^2 \times \mathbb{R}_{>0}$ where $\mathbb{R}^2$ parameterizes the center and $\mathbb{R}_{>0}$ parameterizes the radius. The moduli space of rectangles in $\mathbb{R}^2$ up to translation and rotation is $\{(a, b) \in \mathbb{R}_{>0}^2 \,|\, a \geq b\}$. Often moduli spaces come with natural topologies and even algebraic structures if the collection consists of algebraic objects.

Now a curve in $\mathbb{P}^2$ of degree d is defined as the zero locus

$$\{[z_0, z_1, z_2] \in \mathbb{P}^2 \,|\, f(z_0, z_1, z_2) = 0\} \tag{16}$$

---

[20] The automorphism group $PGL_2 = SL_2/\{\pm 1\}$ of $\mathbb{P}^1$ acts on $\mathbb{P} M^{(d+1) \times (n+1)}$ as $\mathbb{C}[t, s]_d$ is an irreducible representation of SL_2 of dimension $d + 1$.

of a nonzero homogeneous polynomial $f(z_0, z_1, z_2) \in \mathbb{C}[z_0, z_1, z_2]_d$ which is uniquely determined by the curve up to constant. Hence the moduli space of all plane curves of degree d is

$$\mathbb{C}^* \backslash (\mathbb{C}[z_0, z_1, z_2]_d - 0) = \mathbb{P}^{d(d+3)/2}.$$

For a nonzero homogeneous polynomial $f(z_0, z_1, z_2)$, we can detect its degree by restricting it to a general line in $\mathbb{P}^2$ and then finding the number of zeros. The irreducible rational curve in $\mathbb{P}^2$ of degree d defined by a polynomial map (13) for $n = 2$ intersects with a line at d points obviously and hence the moduli space $R_d(\mathbb{P}^2)$ of all irreducible rational curves is a subset of the space $\mathbb{P}^{d(d+3)/2}$ of all plane curves of degree d. In this section, a *rational curve* in $\mathbb{P}^2$ of degree d is defined as a curve in $\mathbb{P}^2$ of degree d in the closure $\overline{R}_d(\mathbb{P}^2)$ in $\mathbb{P}^{d(d+3)/2}$ of $R_d(\mathbb{P}^2)$. In other words, a rational plane curve of degree d is a limit of irreducible rational curves of degree d in $\mathbb{P}^2$. For instance, all plane curves of degree $d \leq 2$ are rational curves and we will see that a plane curve of degree 3 is rational if and only if it is singular.

A plane curve (16) passes through a point $[z_0, z_1, z_2] \in \mathbb{P}^2$ if and only if $f(z_0, z_1, z_2) = 0$, which is a linear equation in the coefficients a_{ijk} of

$$f(z_0, z_1, z_2) = \sum_{i+j+k=d} a_{ijk} z_0^i z_1^j z_2^k.$$

Hence, the set of plane curves of degree d passing through a given point is a hyperplane. As the moduli space of rational curves of degree d in plane is $3d - 1$ dimensional, we may expect that there are only finitely many rational curves in $\mathbb{P}^2$ of degree d passing through $3d - 1$ points.[21] So we may ask the following.

Question 7 How many rational curves in plane of degree d pass through given $3d - 1$ general points? □

Let us denote the answer by N_d for $d \geq 1$. For $d = 1$, we are looking for lines through 2 given distinct points. Of course the number is $N_1 = 1$. For $d = 2$, we are looking for conics through 5 general points. The space of conics in $\mathbb{P}^2$ is the space of coefficients

$$\mathbb{P}^5 = \{[a_0, a_1, \cdots, a_5]\}$$

of quadratic polynomials

$$a_0 z_0^2 + a_1 z_0 z_1 + a_2 z_1^2 + a_3 z_0 z_2 + a_4 z_1 z_2 + a_5 z_2^2 \tag{17}$$

in the homogeneous coordinates z_0, z_1, z_2 of $\mathbb{P}^2$, up to constant. The locus of conics through a given point is given by a linear polynomial in $a_0, \cdots, a_5$. Hence the locus of conics through 5 general points is the intersection of 5 hyperplanes in $\mathbb{P}^5$. For example, we find that there is precisely one conic ($z_0^2 + z_1^2 = z_2^2$) passing through

---

[21] In algebraic terms, we are asking for the degree of the projective variety $\overline{R}_d(\mathbb{P}^2)$ in $\mathbb{P}^{d(d+3)/2}$.

$$[1, 0, 1], [0, 1, 1], [1, 0, -1], [0, 1, -1], [1, \sqrt{-1}, 0]$$

by plugging in these coordinates to (17). By Question 4, we find that $N_2 = 1$.

For $d = 3$, it is an elementary exercise[22] to show that any irreducible homogeneous cubic polynomial in $z_0 = x$, $z_1 = y$, $z_2 = z$ is transformed to

$$y^2 z = x(x - \mu z)(x - \lambda z), \quad \lambda, \mu \in \mathbb{C} \tag{18}$$

by a linear change of coordinates, i.e $z_i' = \sum_j a_{ij} z_j$ for an $A = (a_{ij}) \in GL_3$. It is easy to see that the cubic curve defined by (18) is smooth if and only if λ, μ, 0 are all distinct, in which case the cubic curve is not rational.[23]

When singular, a plane cubic defined by (18) is transformed to either $y^2 z = x^3$ (cusp) or $y^2 z = x^2(x - z)$ (node) by a linear change of coordinates. In both cases, the singular cubic is a rational curve because it is the image of the polynomial map

$$\mathbb{P}^1 \to \mathbb{P}^2, \quad [t, s] \mapsto [t^2 s, t^3, s^3] \quad \text{(cusp)}, \quad \text{or}$$

$$\mathbb{P}^1 \to \mathbb{P}^2, \quad [t, s] \mapsto [s(t^2 + s^2), t(t^2 + s^2), s^3] \quad \text{(node)}.$$

Moreover, all reducible cubic polynomials define rational curves. Therefore an irreducible plane cubic curve is rational if and only if it is singular. Hence, Question 7 for $d = 3$ is the same as the following.

Question 8 How many singular cubic curves in $\mathbb{P}^2$ pass through 8 given general points? □

The locus of cubics passing through a point is a hyperplane in the moduli space $\mathbb{P}^9$ of all plane cubic curves. Hence the locus of cubics passing through 8 general points is a line $\mathbb{P}^1 = \{[t, s]\}$, which means that we have two cubic polynomials $F, G \in \mathbb{C}[z_0, z_1, z_2]_3$ such that a cubic curve passing through 8 general points is the zero locus $C_{[t,s]}$ of $tF + sG$ for some $[t, s] \in \mathbb{P}^1$. A singular point of $C_{[t,s]}$ is characterized by the vanishing of the three partial derivatives

$$t\frac{\partial F}{\partial z_0} + s\frac{\partial G}{\partial z_0}, \quad t\frac{\partial F}{\partial z_1} + s\frac{\partial G}{\partial z_1}, \quad t\frac{\partial F}{\partial z_2} + s\frac{\partial G}{\partial z_2}, \tag{19}$$

---

[22] An irreducible cubic curve $C = (f(x, y, z) = 0)$ has at most one singular point because if there are two, the line joining them meets the cubic curve at two points with multiplicity ≥ 2, which is impossible. A point $[x_0, y_0, z_0] \in C$ is an inflection point of C if the Hessian determinant of f at (x_0, y_0, z_0) is zero. As the Hessian is a polynomial of degree 3, there are 9 inflection points and we may pick a smooth inflection point. By a coordinate change, we may assume the inflection point is $[0, 1, 0]$. By simple algebra, we can choose coordinates such that the irreducible cubic polynomial f is of the form (18).

[23] The smooth cubic plane curve defined by (18) is a double cover of $\mathbb{P}^1$ branched at 4 points. Hence it is topologically a torus $S^1 \times S^1$ which does not allow a ramified covering by $\mathbb{P}^1$, homeomorphic to S^2, by the Hurwitz formula.

which are homogeneous of degree 1 in t, s and of degree 2 in z_0, z_1, z_2. Such a system of three equations has 12 common solutions.[24] For example, let

$$F(x, y, z) = y^2 z - x^3 + x^2 z, \quad G(x, y, z) = x^3 + y^3 + z^3,$$

with $x = z_0$, $y = z_1$, $z = z_2$, whose zero loci are denoted by C and C' respectively. Then one can check by hand that $C \cap C'$ consists of 9 distinct points and a cubic curve passing through 8 points among $C \cap C'$ is the zero locus $C_{[t,s]}$ of $tF + sG$ for some $[t, s] \in \mathbb{P}^1$. It is straightforward to check from the vanishing of (19) that there are exactly 12 singular curves $C_{[t,s]}$ in this family, each of which contains exactly one singular point. Hence the answer to Question 8 is $N_3 = 12$, as was computed by Steiner in 1848.

How about N_d for $d > 3$? The cubic case was considerably more difficult than $d \leq 2$ and you may expect that the problem will become much more difficult as d increases. In fact, between 1848 and 1993 CE, only $N_4 = 620$ and $N_5 = 87304$ were computed. So it was taken as a complete surprise when Kontsevich proved the formula

$$N_d = \sum_{d_1 + d_2 = d} N_{d_1} N_{d_2} d_1^2 d_2 \left(d_2 \binom{3d-4}{3d_1-2} - d_1 \binom{3d-4}{3d_1-1} \right) \qquad (20)$$

in 1994. With the input $N_1 = N_2 = 1$, (20) enables us to compute all N_d inductively.

In 1990s, string theory, especially the mirror symmetry, challenged mathematicians with conjectures on enumeration of curves and Kontsevich formulated the notion of stable maps for a mathematical theory of the Gromov-Witten invariant. Actually, once equipped with the notion of stable maps, Question 7 is not so hard any more. The formula (20) is the first major success story in enumerative geometry motivated by string theory, and we will see a sketchy proof of (20) in Sect. 4.1.

---

[24] Each of the three partial derivatives is a section of the line bundle $\mathcal{O}_{\mathbb{P}^1 \times \mathbb{P}^2}(1, 2)$ over $\mathbb{P}^1 \times \mathbb{P}^2$. Their common zero locus has 12 points by

$$\int_{\mathbb{P}^1 \times \mathbb{P}^2} c_1(\mathcal{O}_{\mathbb{P}^1 \times \mathbb{P}^2}(1, 2))^3 = (p + 2\ell)^3 = p^3 + 6p^2\ell + 12p\ell^2 + \ell^3 = 12p\ell^2 = 12 \in H_0(\mathbb{P}^1 \times \mathbb{P}^2)$$

where ℓ is the class of a line in $\mathbb{P}^2$ and p is the class of a point in $\mathbb{P}^1$ since $p^2 = 0$, $\ell^3 = 0$. As there are 12 singular points in

$$\{([t, s], [z_0, z_1, z_2]) \in \mathbb{P}^1 \times \mathbb{P}^2 \mid tF(z_0, z_1, z_2) + sG(z_0, z_1, z_2) = 0\},$$

there are 12 singular curves for general choices of 8 points, as one can check with explicit examples.

2.3   Enumerating Rational Curves in Hypersurfaces

Another classical enumerative geometry problem is about rational curves in hypersurfaces. Let $Y \subset \mathbb{P}^n$ be a smooth hypersurface[25] defined by the vanishing of a homogeneous polynomial $F(z_0, \cdots , z_n)$ of degree k.

A polynomial map (13) is a map to Y if and only if

$$F(f_0(t, s),\, f_1(t, s), \cdots ,\, f_n(t, s)) = 0 \in \mathbb{C}[t, s]_{dk}. \tag{21}$$

By (14), we find that (21) is equivalent to the vanishing of $dk + 1$ homogeneous polynomials in the coefficients a_{ij}. Let W be the subvariety of $M^{(d+1)\times(n+1)}$ defined by the vanishing of these $dk + 1$ homogeneous polynomials, whose expected dimension is $(d + 1)(n + 1) - dk - 1$. Then the moduli space $R_d(Y)$ of irreducible rational curves of degree d in Y is an open subset in the orbit space

$$SL_2 \backslash \mathbb{P}W \subset SL_2 \backslash \mathbb{P}M^{(d+1)\times(n+1)}$$

whose expected dimension is

$$(d + 1)(n + 1) - dk - 5. \tag{22}$$

It is easy to see that the actual dimension is not smaller than the expected dimension because an equation can drop the dimension by at most 1.

Let us enumerate lines in Y first. As $d = 1$, the expected dimension is $2n - 3 - k$ and hence if $k < 2n - 3$, we find that there are infinitely many lines in the hypersurface Y of degree k. For $k > 2n - 3$, no lines are expected in Y because the expected dimension is negative. So we ask the following.

Question 9 How many lines lie in a smooth hypersurface Y of degree $2n - 3$ in $\mathbb{P}^n$ for $n \geq 2$?

For $n = 2$, the answer is 1 obviously. For $n = 3$, $2n - 3 = 3$ and Y is a cubic surface in $\mathbb{P}^3$. The following is a beautiful result which you can find in an undergraduate textbook on algebraic geometry.

Theorem 1 *Every smooth cubic surface in $\mathbb{P}^3$ has exactly 27 lines.*

There are many ways to prove this theorem as follows:

1. Check that the number of lines is independent of the cubic surface and enumerate lines in a special cubic surface like the Fermat cubic

---

[25] A hypersurface Y defined by a polynomial $F \in \mathbb{C}[z_0, \cdots , z_n]$ is smooth if and only if the differential

$$dF = \frac{\partial F}{\partial z_0} dz_0 + \cdots + \frac{\partial F}{\partial z_n} dz_n$$

is nonzero at every point of Y.

$$z_0^3 + z_1^3 + z_2^3 + z_3^3 = 0$$

explicitly. For instance, the line defined by $z_0 + e^{\frac{2\pi\sqrt{-1}}{3}} z_1 = 0 = z_2 + z_3$ is contained in the Fermat cubic surface. Did you get 27?

2. Show that every smooth cubic surface is the blowup of $\mathbb{P}^2$ at 6 points $q_1, \cdots, q_6$.[26] The lines in Y are obtained from 15 lines in $\mathbb{P}^2$ through 2 points among $\{q_i\}$, 6 conics in $\mathbb{P}^2$ through 5 points among $\{q_i\}$ and 6 exceptional divisors. Hence $27 = 15 + 6 + 6$ lines.

3. Show that the lines in Y are all rigid, i.e. they do not deform in Y. As (21) is a section of the vector bundle $E = \pi_* \mathcal{O}_{\mathbb{P}\mathcal{U}}(3)$ of rank 4 over the Grassmannian $Gr(2, 4)$ of lines in $\mathbb{P}^3$, the number of points in its vanishing locus is computed by the Euler class[27] of E, where $\mathcal{U}$ is the universal rank 2 bundle over $Gr(2, 4)$ and $\pi : \mathbb{P}\mathcal{U} \to Gr(2, 4)$ is the bundle projection. By Poincaré-Hopf and Riemann-Roch, we find the number of lines in Y to be

$$\int_{Gr(2,4)} c_4(\pi_* \mathcal{O}_{\mathbb{P}\mathcal{U}}(3)) = 27.$$

More details of the proofs and related geometry can be found in most textbooks on algebraic geometry.

The next case for Question 9 is when $n = 4$ and hence $k = 2n - 3 = 5$. This case is quite special in that the expected dimension (22) vanishes for all d. So we expect a finite number of rational curves of degree d in a smooth quintic hypersurface in $\mathbb{P}^4$ for any $d > 0$. Therefore a complementary question to Question 9 is the following.

---

[26] The linear system $3H - E_1 - \cdots - E_6$ gives an embedding of the blowup of $\mathbb{P}^2$ at 6 points into $\mathbb{P}^3$ as a cubic surface, where E_i are the exceptional divisors. The moduli of cubic surfaces is $4 = \binom{6}{3} - 4^2$ dimensional while the choice of 6 points modulo automorphism of $\mathbb{P}^2$ has $4 = 6 \cdot 2 - (3^2 - 1)$ moduli. One can further compare the deformation theories.

[27] A complex vector bundle E of rank r over a topological space X is a continuous map $\pi : E \to X$ of topological spaces such that there is an open cover $X = \cup_\alpha X_\alpha$ with a homeomorphism $\varphi_\alpha : \pi^{-1}(X_\alpha) \to X_\alpha \times \mathbb{C}^r$ for each α such that

$$\varphi_\beta \circ \varphi_\alpha^{-1} : (X_\alpha \cap X_\beta) \times \mathbb{C}^r \to (X_\alpha \cap X_\beta) \times \mathbb{C}^r, \quad (x, v) \mapsto (x, g(x)(v))$$

for a continuous $g : X_\alpha \cap X_\beta \to GL_r(\mathbb{C})$. A section of the vector bundle $\pi : E \to X$ is a continuous map $s : X \to E$ satisfying $\pi \circ s = \mathrm{id}_X$. For all $x \in X$, $\pi^{-1}(x)$ is a vector space and the assignment $x \mapsto 0 \in \pi^{-1}(x)$ is a continuous map $0 : X \to E$ satisfying $\pi \circ 0 = \mathrm{id}_X$, called the zero section. It is obvious that the zero section is a homeomorphism onto its image and hence we can think of X as a subspace of E by the inclusion $0 : X \to E$. Under reasonable assumptions like X being a CW complex, it is not hard to see that for any cycle ξ in X we have a (perturbed zero) section which is transversal to ξ. The Euler class of E is a homomorphism

$$c_r(E) : H_*(X) \longrightarrow H_{*-2r}(X), \quad \xi \mapsto c_r(E) \cap \xi = \int_\xi c_r(E)$$

sending ξ to the intersection of ξ with a section of E, transversal to ξ. If X is a complex manifold, $c_r(E)$ is represented by a differential form of degree $2r$, supported in a small neighborhood of the zero section.

Question 10 How many rational curves of degree $d \geq 1$ lie in a smooth quintic 3-fold Y in $\mathbb{P}^4$? ☐

To answer this question, one first proves that rational curves in Y are rigid.[28] As in the case of cubic surfaces, for $d = 1$, by Poincaré-Hopf, the answer to Question 10 is the integral

$$\int_{Gr(2,5)} c_6(\pi_* \mathcal{O}_{\mathbb{P}\mathcal{U}}(5)) \tag{23}$$

of the Euler class of the rank 6 vector bundle $\pi_* \mathcal{O}_{\mathbb{P}\mathcal{U}}(5)$ over the Grassmannian $Gr(2, 5)$ of lines in $\mathbb{P}^4$. Here the notation is similar to the cubic surface case above. Once again it is straightforward to compute (23) by Riemann-Roch. The answer for $d = 1$ is 2875.

Similar computations with the moduli spaces of conics and twisted cubics tell us that the answer to Question 10 is 609250 for $d = 2$, 317206375 for $d = 3$. The integral like (23) for $d \geq 3$ is much more tricky because we need to compactify the moduli space $R_d(\mathbb{P}^4)$ of irreducible rational curves and a natural compactification such as the Hilbert scheme is often not smooth. So the computations were very difficult and algebraic geometers couldn't make much progress on Question 10.

This was more or less the state of knowledge at around 1990 when a group of string theorists found a (conjectural) answer to Question 10 for all d. Certainly, algebraic geometers were completely taken by surprise and this challenge by string theory has nourished algebraic geometry enormously during the past three decades.

2.4 Moduli Space and Intersection Theory

Methods employed in Sect. 2.1 work for many problems about lines, planes and hyperplanes. Answering enumerative geometry problems on them requires finding intersection numbers of subsets in the corresponding moduli spaces of linear objects, called the *Grassmannians*.

Schubert calculated various intersection numbers on Grassmannians, ingeniously but sometimes not so rigorously. In 1900, Hilbert included in his famous list of 23 problems for the 20th century mathematicians the problem of providing a rigorous foundation for Schubert's calculus. It seems fair to say that such a foundation was provided in 1978 by Fulton and MacPherson.

---

[28] More precisely, we need $H^0(N_{C/Y}) = 0$ for any irreducible rational curve C in Y. The rigidity was proved by Sheldon Katz in 1986 for $d \leq 7$. It is not so hard to see directly for $d = 1$. For higher d, Katz first proves that the incidence variety of pairs (C, Y) where Y is a quintic hypersurface and C is a rational curve in Y is irreducible by regularity, for $d \leq 7$. Next he proves that there is at least one smooth rational curve in Y of degree d which is rigid by Mori's argument. Then one can easily conclude that the proper subvariety of rational curves C with $H^0(N_{C/Y}) \neq 0$ is empty for general Y.

The keys for Fulton and MacPherson's intersection theory are the *degeneration to normal cone*[29] and *rational equivalence*.[30] When intersecting a subset Y in a variety X with another subset, we may replace X by the normal cone $C_{Y/X}$ up to rational equivalence by a degeneration, i.e. a continuous family X_t for $t \in \mathbb{C}$ with $X_1 = X$ and $X_0 = C_{Y/X}$. If Y is a smooth subvariety so that the normal cone is in fact the normal bundle

$$N_{Y/X} = T_X|_Y / T_Y,$$

the quotient of the tangent bundle of X by the tangent bundle of Y, then the intersection of Z with Y is the result of intersecting the cone $C_{Y \cap Z/Z} \subset N_{Y/X}$ with the zero section of $N_{Y/X}$.

Intersection with the zero section of a vector bundle is taken care of by the Euler class of the bundle, and is obtained by either (1) perturbing the zero section until it becomes transversal to the cone $C_{Y \cap Z/Z}$ or (2) perturbing the cone to make it transversal to the zero section. Usually the first perturbation is preferred in topology but in algebraic geometry the second perturbation has to be used because often the zero section is impossible to perturb.

Even when the normal cone $C_{Y/X}$ is not a vector bundle, if we can embed it into a vector bundle E, a similar argument works but then the intersection theory depends on the embedding $\iota : C_{Y/X} \to E$. So if you can find E and ι in a natural fashion from geometric considerations, we can still perform intersections of subvarieties. Modern enumerative geometry since 1995 is based on this observation, under the name of *virtual intersection theory* that we will see below.

We found the answer to Question 6 in two ways and in fact these are the main roads in enumerative geometry. Let us recapitulate them. The first way to answer Question 1 proceeds as follows:

1. Construct the moduli space[31] of all geometric figures of fixed type (like $Gr(2, 4)$);
2. Find the loci of geometric figures satisfying the given constraints (like the hyperplanes);
3. Find the intersection of the loci.

It is usually impossible to find the intersection points of subsets explicitly, but in case there are only finitely many intersection points, their number is often expressed in

---

[29] The normal cone is obtained by considering the lowest order terms of the defining polynomials. For instance, the normal cone of the singular cubic curve $y^2 = x^2(x + 1)$ (resp. $y^2 = x^3$) at the origin is the union $y^2 = x^2$ of the two tangent lines (resp. the double line $y^2 = 0$).

[30] Two irreducible subvarieties X, Y of dimension d in a variety V are rationally equivalent if there is a $d + 1$ dimensional subvariety W of V and a rational function on W whose zero is X and pole is Y. This relation gives us the rational equivalence on the free group generated by irreducible subvarieties.

[31] As we saw above, moduli spaces are often constructed as orbit spaces under group actions. The art of finding nice orbit spaces or closest geometric objects belonged to the realm of *geometric invariant theory* (GIT) but now the method of *algebraic stacks* is as useful as GIT. The moduli spaces often depend on choices and they undergo suitable transformations as we vary our choices. The changes in the answers to enumerative problems as we vary the choices are called *wall crossing* terms.

terms of cohomology pairings and hence integrals of differential forms (by Poincaré duality). Therefore, enumerative geometry is mostly about constructing **moduli spaces** and **intersection theory** of subvarieties.[32]

The second way is to deform the problem continuously till you reach a more tractable situation. But you need to prove ahead that the answer remains invariant (unchanged) during the deformation. So it proceeds as follows:

1. Prove that the answer is deformation invariant;
2. Solve the special case.

The technique of deformation (or degeneration in case we allow singularities) is very important in classical and modern enumerative geometry.

3 Enter String Theory

So far, we've seen a few snapshots of classical enumerative geometry. Throughout history, mathematics has been enriched by interactions with other disciplines like philosophy, astronomy, navigation, gambling, artillery, physics and engineering. Quite recently, enumerative geometry has been considerably expanded by interacting with string theory since 1990s. In this section, I want to discuss why the interaction was possible and how it happened. But I have to keep the discussion minimal because my knowledge is quite limited.

Newton's mechanics was founded on Euclidean geometry. By calculus, he proved that when dealing with a solid body in mechanics, we may assume that the object is a point located at its center of mass. So Newton's physics is about points and their trajectories. Even today, the standard physics explains elementary particles as points. To a geometer like me, string theory looks like an attempt to make physics more interesting by adding more geometric structures to physics. Thanks to string theory, we can now think of elementary particles as curves, surfaces and so on. The price we pay is that to make the theory consistent with our universe, we have to accept the assumption that the universe has extra dimensions, on top of the usual 4 dimensional space-time.

The use of extra dimension dates back to 1919 when Kaluza observed that if we increase the dimension of the space-time $\mathbb{R}^4$ by adding a tiny circle S^1, then Einstein's general relativity and Maxwell's electromagnetism are amalgamated into a concise and united theory. Around 1985, string theorists including Witten realized that the

---

[32] Intersection theory is the study of finding the equivalence class of the intersection of suitable subsets. Depending on the objects and the equivalence relation, there are different layers of intersection theories, like Chow theory, K-theory, Borel-Moore homology, algebraic cobordism, motivic cohomology etc, and Riemann-Roch compares them.

correct way to enlarge the space-time in string theory is to add a Calabi-Yau 3-fold.[33] Perhaps the simplest example of a compact Calabi-Yau 3-fold is a smooth quintic hypersurface in $\mathbb{P}^4$ like the Fermat quintic

$$Y = (z_0^5 + z_1^5 + \cdots + z_4^5 = 0) \subset \mathbb{P}^4.$$

So now all the physics takes place in $\mathbb{R}^4 \times Y$ for a Calabi-Yau 3-fold Y. When a particle is a point, its trajectory is a curve. But when a particle is a string like S^1, its trajectory (world sheet) is a map from a surface $S^1 \times \mathbb{R}$ to $\mathbb{R}^4 \times Y$. By projecting to Y, we obtain a map $S^1 \times \mathbb{R} \to Y$. The length of the string may change from 0 to R and then to 0. Also strings may ramify into two strings or more. Hence we think of a world sheet as a map $f : \Sigma \to Y$ from a Riemann surface Σ. If we choose a Lagrangian suitably, the Euler-Lagrange equation is satisfied at f if and only if f is holomorphic, i.e. $f(\Sigma)$ is an algebraic curve in case Y is contained in a projective space. So enumerating curves in the Calabi-Yau 3-fold Y may be interpreted as an integral on the critical set of the action functional, a physics quantity in string theory. This observation connects string theory with enumerative geometry.

There are two ways to obtain Lagrangians relevant to enumerative geometry, A-model and B-model. An interesting proposition in string theory says that for a Calabi-Yau 3-fold Y, there is a mirror Calabi-Yau 3-fold $\hat{Y}$ such that the A-model physics of Y (resp. $\hat{Y}$) is equivalent to the B-model physics of $\hat{Y}$ (resp. Y). This is called the mirror symmetry.

Around 1990, it occurred to a group of string theorists that the mirror symmetry may be used to solve enumerative geometry problems. The A-model physics of a quintic 3-fold Y in $\mathbb{P}^4$ enumerates rational curves in Y while the B-model of the mirror partner $\hat{Y}$ gives us an explicit (hypergeometric) series. So the physics of mirror symmetry predicted the number of rational curves of degree d in Y for all d. It was a huge blow to algebraic geometers because string theory provided all the numbers while algebraic geometry was not making progress towards Question 10. Since then, many conjectural formulas on enumerative geometry problems have been produced by applying string theory and interesting new developments followed through attempts to rigorously prove or disprove the conjectures in mathematics.

4   Modern Enumerative Geometry

In this section, let us take a look at new aspects of enumerative geometry since the interaction with string theory.

As we saw in Sect. 2, a standard approach to enumerative geometry problems consists of

---

[33] A 3-fold is a Kahler manifold of dimension 3. A complex manifold Y of dimension r is *Calabi-Yau* if there is a nowhere vanishing holomorphic r-form on Y, i.e. the canonical line bundle $K_Y = \det T_Y^* \cong \mathcal{O}_Y$ is trivial.

1. construction of a moduli space,
2. intersection theory of subvarieties and
3. checking that the numbers are actually enumerating the objects we want.

Let us not worry about the issue (3) and focus on getting a number by (1) and (2). When the moduli space is smooth, by Poincaré duality,[34] intersection theory of subvarieties is taken care of by integrals of closed differential forms η. We call an integral $\int_X \eta$ of a differential form η (or a cohomology class) against the fundamental class[35] of a moduli space X, an *enumerative invariant*, if the number is fixed under deformation of the enumerative problem. As noted in Sect. 2, deformation to a simpler case is an extremely valuable technique.

In reality, moduli spaces we will use are highly singular and an integral of the form $\int_X \eta$ will never be deformation invariant, even if we could make sense of it. The way out is to find a replacement of the fundamental class, namely a homology class of expected dimension which is invariant under deformation. The replacement is now called a *virtual fundamental class* and denoted by $[X]^{\mathrm{vir}}$. So the issue (2) should be split into two steps of finding a virtual fundamental class $[X]^{\mathrm{vir}}$ and then integrating cohomology classes against $[X]^{\mathrm{vir}}$. In summary, designing an enumerative invariant requires three steps:

Step 1. Construct a nice moduli space X;
Step 2. Construct a virtual fundamental class $[X]^{\mathrm{vir}}$;
Step 3. Integrate cohomology classes against $[X]^{\mathrm{vir}}$.

Let us see how each step works through the example of *Gromov-Witten invariant*. Unfortunately, the discussions from now on will get more technical, sometimes beyond the level of undergraduate mathematics.

4.1 Step 1: Moduli Space

As we saw in Sect. 3, the motivation for modern enumerative invariants came from the mirror symmetry. The first task was to make clear the meaning of the numbers calculated by the mirror symmetry. Mathematicians realized very quickly that the numbers enumerate symplectic surfaces in the symplectic manifold underlying the quintic 3-fold Q but not as embedded symplectic manifolds but as maps.[36] In

---

[34] The Poincaré duality identifies homology classes (submanifolds without boundary modulo homological equivalence) with cohomology classes (closed differential forms modulo exact forms).

[35] The fundamental class of an oriented manifold is the whole space with the chosen orientation. To make sense of an integral, we need an orientation and a change of orientation may change the sign componentwise.

[36] A mathematical count of embedded symplectic surfaces Σ in a symplectic manifold Q is called the Gromov invariant, which equals the Seiberg-Witten invariant when Q is a symplectic 4-manifold. When the 4-manifold is a smooth projective surface, the Gromov/Seiberg-Witten invariant is the virtual integral (called the Poincaré invariant) on the Hilbert scheme of curves and is equivalent to the Donaldson invariant (after wall crossing) by Mochizuki's formula.

algebraic setting, the numbers from the mirror symmetry enumerate polynomial (regular) maps $f : C \to Q$, up to automorphism, where C is a projective algebraic curve, with at worst nodal singularities.[37] Such a map f is called *stable* if the automorphism group

$$\mathrm{Aut}(f) = \{\varphi : C \xrightarrow{\cong} C \mid f \circ \varphi = f\}$$

of invertible polynomial maps of C preserving f is finite.

How do we construct the moduli space of stable maps to a smooth projective variety Q? There are several approaches: One may use geometric invariant theory starting from a big Hilbert scheme, or one may show that the moduli functor is an algebraic stack and then prove its expected properties like projectivity. A more down-to-earth construction due to Fulton and Pandharipande goes as follows: Let us denote by $\overline{M}_{g,n}(Q, d)$ the moduli space of stable maps $f : C \to Q \subset \mathbb{P}^r$ from curves C of genus g with n smooth distinct marked points such that the direct image $f_*[C]$ of the fundamental class of C meets a general hyperplane at d points. A polynomial map $f : C \to \mathbb{P}^r$, $f(p) = [f_0(p), \cdots, f_r(p)]$ is determined by its component functions f_i. Each polynomial function f_i is determined by its d roots up to constant. For a stable map f, we can choose a coordinate system for $\mathbb{P}^r$ such that all these roots are distinct and away from the marked and singular points of C. Hence the roots determine a pointed curve for which we already have a moduli space $\overline{M}_{g,m}/(S_d)^{r+1}$ for $m = n + d(r+1)$ whose dimension is $m + 3g - 3$.[38] The choice of f from the roots requires r parameters[39] and a fiber bundle over an open subspace in $\overline{M}_{g,m}/(S_d)^{r+1}$ gives us a chart of $\overline{M}_{g,n}(\mathbb{P}^r, d)$. Gluing these charts constructs $\overline{M}_{g,n}(\mathbb{P}^r, d)$.[40] If you understand this construction, it is not hard to see that for any projective variety $Q \subset \mathbb{P}^r$, the moduli space of stable maps to Q is a closed subspace

---

[37] Here are the definitions: A projective algebraic curve is a projective variety of dimension 1. A projective variety Q is a subset of a projective space $\mathbb{P}^n$ defined by the vanishing of some homogeneous polynomials $f_1, \cdots, f_m$ in the homogeneous coordinates $x_0, \cdots, x_n$. The dimension of Q is 1 if n minus the rank of the Jacobian of $(f_1, \cdots, f_m)$ is 1 at all points of Q except finitely many points. A projective algebraic curve C has at worst nodal singularity if in an analytic neighborhood of each point $p \in C$, the curve is smooth or looks like the union of two lines ($xy = 0$). Topologically a projective algebraic curve C with at worst nodal singularities is the union of finitely many doughnuts possibly with any number of holes, glued at points. The genus of C is the total number of holes plus the number of loops in the dual graph. When the genus g is zero, we say C is rational.

[38] The moduli space $\overline{M}_{g,n} = \overline{M}_{g,n}(pt, 0)$ of stable curves was constructed by Deligne and Mumford in 1968. Here a stable curve means a stable map to a point. Any subgroup G of the symmetric group S_n acts on $\overline{M}_{g,n}$ by permuting the marked points.

[39] We have to find isomorphisms of the line bundles L_i defined by the Cartier divisor zero(f_i). Since $\mathrm{Hom}(L_i, L_i) = H^0(\mathcal{O}_C)$ is 1 dimensional, the identification of line bundles $L_0, \cdots, L_r$ requires r dimensional parameter space. The cohomology $H^1(\mathcal{O}_C) = \mathbb{C}^g$ is the obstruction space to gluing of line bundles and hence the expected dimension is $m + 3g - 3 + r(1 - g) = n + (3 - r)(g - 1) + d(r + 1)$.

[40] The projectivity follows from Kollár's criterion.

$$\overline{M}_{g,n}(Q, d) \subset \overline{M}_{g,n}(\mathbb{P}^r, d)$$

which is also projective.[41]

As an application of the construction of the moduli space of stable maps, let us prove Kontsevich's formula (20).

Proof **(Sketchy proof of Kontsevich's formula** (20)) We will prove the formula for $d = 3$ (Question 8) and leave the general case to the reader which is no more difficult. Pick 7 general points $q_3, q_4, \cdots, q_9$ and two lines L_1 and L_2 in general position. In the moduli space $\overline{M}_{0,9}(\mathbb{P}^2, 3)$, let Y be the locus of stable maps whose first marked point p_1 maps to a point in L_1, the second marked point p_2 to a point in L_2, and the jth marked point p_j to q_j for $j \geq 3$. The dimension of $\overline{M}_{0,9}(\mathbb{P}^2, 3)$ is 17 and the requirements for p_1 and p_2 cut the dimension by 1 each while the requirement for p_j, $j \geq 3$ drops the dimension by 2 each. Hence you find that Y is a projective curve.

Consider the composition

$$\varphi : Y \subset \overline{M}_{0,9}(\mathbb{P}^2, 3) \longrightarrow \overline{M}_{0,9} \longrightarrow \overline{M}_{0,4} \cong \mathbb{P}^1$$

where the first arrow forgets the map to $\mathbb{P}^2$ and the second arrow forgets the marked points p_j for $j > 4$. By the Cauchy integration theorem or its algebraic counterpart, a meromorphic function on a projective curve has the same number of zeros and poles. Hence, $\varphi^{-1}(0)$ and $\varphi^{-1}(\infty)$ have the same number of points. By change of coordinates if necessary, we may say $0 \in \mathbb{P}^1$ (resp. $\infty \in \mathbb{P}^1$) represents a nodal rational curve

$$C = C_1 \cup C_2 \in \overline{M}_{0,4} \cong \mathbb{P}^1$$

with two components, with p_1, $p_2 \in C_1$ (resp. p_1, $p_3 \in C_1$) and p_3, $p_4 \in C_2$ (resp. p_2, $p_4 \in C_2$).

Now let us count the number of stable maps $f \in \varphi^{-1}(0)$.

1. If the degree of f restricted to the component C_1 is 0, then this component should map to $L_1 \cap L_2$ and the image of f passes through all the eight points $L_1 \cap L_2, q_3, \cdots, q_9$. Hence there are N_3 such stable maps.
2. If the degree of f restricted to C_1 is 1, then the other component C_2 has degree 2 and hence $f(C_2)$ may contain at most 3 points out of $q_5, \cdots, q_9$ because $f(C_2)$ already passes through q_3 and q_4. Hence $f(C_1)$ should pass through exactly 2 points out of $q_5, \cdots, q_9$. We have $\binom{5}{2} = 10$ choices of the two points and the gluing point of C_1 and C_2 may be mapped to one of the two points of the intersection of the line by C_1 and the conic by C_2. Hence there are 10×2 such stable maps.
3. If the degree of f restricted to C_1 is 2, the line $f(C_2)$ has to be the line joining q_3 and q_4. Thus $f(C_1)$ is the unique conic through $q_5, \cdots, q_9$. As $f(C_1) \cap L_1$ and

---

[41] To be precise, the Fulton-Pandharipande construction gives us the Deligne-Mumford stack which represents the natural moduli functor of stable maps whose coarse moduli space is a projective scheme. Coarse moduli means that we are forgetting the stabilizer groups of points.

$f(C_1) \cap L_2$ have two points each, there are 2×2 choices for $f(p_1)$ and $f(p_2)$. As the conic $f(C_1)$ meets the line $f(C_2)$ at two points, we have two choices for $f(C_1 \cap C_2)$. Hence there are $2 \times 2 \times 2$ such stable maps.

In summary, there are $N_3 + 20 + 8$ points in $\varphi^{-1}(0)$.

Let us count the number of stable maps $f \in \varphi^{-1}(\infty)$.

1. The degree of f restricted to the component C_1 cannot be 0 or 3 because $q_3 \notin L_1$ and $q_4 \notin L_2$.
2. If the degree of f restricted to C_1 is 1, $f(C_1)$ has to be one of the five lines joining q_3 with one of $q_5, \cdots, q_9$ while $f(C_2)$ is the unique conic passing through q_4 and the remaining 4 points among $q_5, \cdots, q_9$. As $f(C_2)$ meets L_2 at 2 points, there are 2 choices for $f(p_2)$. As the line $f(C_1)$ meets the conic $f(C_2)$ at 2 points, there are two choice for $f(C_1 \cap C_2)$. Hence there are $5 \times 2 \times 2$ such stable maps.
3. If the degree of f restricted to C_1 is 2, the same argument tells us that there are 20 such stable maps.

In summary there are $20 + 20 = 40$ points in $\varphi^{-1}(\infty)$.

We thus find that $N_3 + 28 = 40$ and hence $N_3 = 12$ as we saw in Sect. 2.2.   □

4.2   Step 2: Virtual Fundamental Class

To discuss anything meaningful, we absolutely need the notion of a vector bundle. Let us begin with the tangent bundle and define a vector bundle as a generalization.

For an open $U \subset \mathbb{R}^n$, tangent vectors are expressions like

$$a = a_1 \frac{\partial}{\partial x_1} + \cdots + a_n \frac{\partial}{\partial x_n}, \quad a_i \in \mathbb{R} \tag{24}$$

and if f is a smooth function on U, its derivative in the direction of a is

$$df(a) = \sum_i a_i \frac{\partial f}{\partial x_i}.$$

By the assignment $a \mapsto (a_1, \cdots, a_n) \in \mathbb{R}^n$, you may think of $U \times \mathbb{R}^n$ as the set of all tangent vectors on U and let $\mathrm{pr} : U \times \mathbb{R}^n \to U$ be the projection to the first component. A vector field on U is then a map $a : U \to U \times \mathbb{R}^n$ whose composition with pr is the identity map on U. For a map $\pi : X \to Y$, a map $s : Y \to X$ with $\pi \circ s = \mathrm{id}_Y$ is often called a *section* of π. For instance, $0 : U \to U \times \mathbb{R}^n$, $0(p) = (p, 0)$ is a section of $\mathrm{pr} : U \times \mathbb{R}^n \to U$, called the *zero section*.

However this description of tangent vectors and vector fields depends on the choice of a coordinate system. If we employ a new coordinate system $\{y_1, \cdots, y_n\}$ instead of $\{x_1, \cdots, x_n\}$, then the expression (24) is transformed to

$$a = \sum_{i=1}^{n} a_i' \frac{\partial}{\partial y_i}, \quad \text{where } a_i' = \sum_{j=1}^{n} a_j \frac{\partial y_i}{\partial x_j}. \tag{25}$$

Hence the identification of the set of all tangent vectors on U with $U \times \mathbb{R}^n$ requires the transformation

$$U \times \mathbb{R}^n \longrightarrow U \times \mathbb{R}^n, \quad (p, a) \mapsto (p, Ja)$$

where $J = (\frac{\partial y_i}{\partial x_j})$ is the Jacobian, which is a smooth map $J : U \to GL(n, \mathbb{R})$.

Let M be a smooth manifold of real dimension n. Then we have an open cover $M = \cup_\alpha M_\alpha$ together with an open embedding $\varphi_\alpha : M_\alpha \to \mathbb{R}^n$. By the above paragraph, the set of all tangent vectors on M_α can be identified with $M_\alpha \times \mathbb{R}^n$. On the intersection $M_{\alpha\beta} := M_\alpha \cap M_\beta$, we have two coordinate systems coming from φ_α and φ_β. By the above paragraph, the identifications of the set of tangent vectors on $M_{\alpha\beta}$ are related by the Jacobian $J_{\alpha\beta} : M_{\alpha\beta} \to GL(n, \mathbb{R})$ of $\varphi_\beta \circ \varphi_\alpha^{-1}$. It is easy to see that the identity, called the *2-cocycle condition*,

$$J_{\gamma\alpha} \circ J_{\beta\gamma} \circ J_{\alpha\beta} = \text{id}$$

holds on $M_\alpha \cap M_\beta \cap M_\gamma$ for all indices α, β, γ, by the chain rule. Let us define the *tangent bundle* of M to be the set of equivalence classes

$$T_M = \sqcup_\alpha (M_\alpha \times \mathbb{R}^n)/ \sim$$

where $(p, v) \sim (q, w)$ for $p \in M_\alpha$ and $q \in M_\beta$ if and only if $p = q$ and $w = J_{\alpha\beta}(p)v$. This set comes with the projection $\pi : T_M \to M$ which equals the projection $\text{pr} : M_\alpha \times \mathbb{R}^n \to M_\alpha$ over the open set M_α. A vector field on M is just a section of the tangent bundle $\pi : T_M \to M$. By declaring that $\pi^{-1}(M_\alpha)$ is open in T_M and the obvious map $M_\alpha \times \mathbb{R}^n \to \pi^{-1}(M_\alpha)$ is a homeomorphism for every α, the tangent bundle T_M becomes a topological space and a smooth manifold.

More generally, a *real vector bundle* of rank r over a topological space M is a topological space E and a continuous map $\pi : E \to M$ such that there are an open cover $M = \cup_\alpha M_\alpha$, a commutative diagram

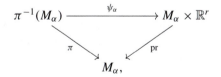

for each α, and a continuous map $\xi_{\alpha\beta} : M_{\alpha\beta} \to GL(r, \mathbb{R})$ for every pair α, β of indices, such that

$$\psi_\beta \circ \psi_\alpha^{-1}(p, v) = (p, \xi_{\alpha\beta}(p)v), \quad \forall p \in M_{\alpha\beta}, v \in \mathbb{R}^r.$$

If we replace continuous (resp. real) by smooth (resp. complex), we obtain a smooth (resp. complex) vector bundle. When M is a complex manifold,[42] a complex vector bundle $\pi : E \to M$ defined by $\xi_{\alpha\beta} : M_{\alpha\beta} \to GL(r, \mathbb{C})$ is a *holomorphic* bundle if $\xi_{\alpha\beta}$ is a holomorphic map. For a holomorphic bundle $\pi : E \to M$, E is a complex manifold. A section $s : U \to E$ on an open $U \subset M$ of a holomorphic bundle $\pi : E \to M$ is holomorphic if it is a holomorphic map of complex manifolds. For a vector bundle $\pi : E \to M$ and a subset N of M, $\pi^{-1}(N)$ is denoted by $E|_N$ and the restriction $E|_N \to N$ of π is a vector bundle over N.

Given a vector bundle E of rank r over M and a homomorphism $\rho : GL(r, \mathbb{R}) \to GL(k, \mathbb{R})$, we obtain a new vector bundle of rank k on M by the recipe

$$\rho(E) := \sqcup_\alpha (M_\alpha \times \mathbb{R}^k)/ \sim$$

where $(p, v) \sim (q, w)$ for $p \in M_\alpha, q \in M_\beta$ if and only if $p = q$ and $w = \rho(\xi_{\alpha\beta}(p))v$. For instance, the determinant $\det : GL(r, \mathbb{R}) \to GL(1, \mathbb{R})$ defines the determinant line bundle $\det(E)$ for a vector bundle E of rank r on M. Here a line bundle means a vector bundle of rank 1. Moreover, given vector bundles E and F, applying linear algebra operations to $\xi_{\alpha\beta}$, we can generate new vector bundles $E \oplus F$, $E \otimes F$, E^*, $\wedge^i E$, $\text{sym}^i E$ and so on. With a help of functional analysis, one can extend the theory of vector bundles to infinite rank vector bundles over infinite dimensional Banach manifolds.

Many moduli spaces in geometry are defined as the zero locus of a smooth section s of an *infinite dimensional* vector bundle $\mathcal{E}$ over an *infinite dimensional* (Banach) manifold $\mathcal{Y}$, at least locally. For instance, if we are looking for the moduli space of holomorphic maps $f : C \to Q$ from a compact Riemann surface[43] C to a smooth projective variety Q, we consider the space $\mathcal{Y}$ of smooth maps $f : C \to Q$ with fixed $f_*[C] \in H_2(Q)$. By the existence and uniqueness of solutions of ordinary differential equations, for each $f \in \mathcal{Y}$, the tangent space to $\mathcal{Y}$ at f is the infinite dimensional vector space

$$T_f \mathcal{Y} = \Gamma(C, f^* T_Q)$$

of smooth sections of the pullback bundle

$$f^* T_Q = \{(p, q) \in C \times T_Q \mid \pi(q) = f(p)\} \longrightarrow C, \quad (p, q) \mapsto p,$$

and there is an open neighborhood of f in $\mathcal{Y}$ diffeomorphic to an open neighborhood of 0 in $T_f \mathcal{Y}$. In particular, $\mathcal{Y}$ is an infinite dimensional smooth manifold.

The condition that the smooth map f should be holomorphic is the vanishing of the section $s = df \circ j - J \circ df$ of the infinite dimensional vector bundle $\mathcal{E}$ whose

---

[42] A complex manifold of complex dimension n is a smooth manifold M with an open cover $M = \cup_\alpha M_\alpha$, a homeomorphism $\phi_\alpha : M_\alpha \to U_\alpha \subset \mathbb{C}^n$ for an open set U_α in $\mathbb{C}^n$ such that the transitions $\phi_\beta \circ \phi_\alpha^{-1}$ are holomorphic on $\phi_\alpha(M_{\alpha\beta})$.

[43] A Riemann surface is a complex manifold of dimension 1.

fiber[44] over f is the vector space $\Gamma(C, Hom(T_C, f^*T_Q))$ of smooth sections of the bundle $Hom(T_C, f^*T_Q)$, where j (resp. J) is the complex structure on T_C (resp. T_Q).[45]

Given an infinite dimensional manifold $\mathcal{Y}$ and an infinite dimensional vector bundle $\mathcal{E}$, a section s of $\mathcal{E}$ is called *Fredholm* if the kernel and cokernel of the differential $ds : T_\mathcal{Y} \to \mathcal{E}$ at each $p \in s^{-1}(0)$ are finite dimensional. In many moduli problems in geometry, the differential ds is an elliptic differential operator and hence Fredholm.

We are interested in the geometry of $M = s^{-1}(0)$ where $0 \subset \mathcal{E}$ denotes the zero section. For any compact subset K of M, we can find an open neighborhood $\mathcal{U}$ in $\mathcal{Y}$ of K and a subbundle $\widetilde{F}$ of $\mathcal{E}|_\mathcal{U}$ of finite rank such that

$$\widetilde{F} + \mathrm{im}(ds) = \mathcal{E}|_\mathcal{U}.$$

Then s gives a section $\bar{s}$ of the quotient $\bar{\mathcal{E}} = \mathcal{E}|_\mathcal{U}/\widetilde{F}$ of $\mathcal{E}|_\mathcal{U}$ by $\widetilde{F}$ whose differential $d\bar{s}$ is surjective by our choice of $\widetilde{F}$. By the implicit function theorem, $Y = \bar{s}^{-1}(0)$ is a finite dimensional manifold and s becomes a section of $F := \widetilde{F}|_Y$ whose zero locus is the open subset $U := \mathcal{U} \cap M$ of M. We thus obtain a finite dimensional local model

$$
\begin{array}{cc}
F & \qquad\qquad (26) \\
\Big\uparrow\Big\downarrow{}_{s} & \\
U = s^{-1}(0) \longrightarrow Y &
\end{array}
$$

for the zero locus M of a Fredholm section, where Y is a finite dimensional manifold and $F \to Y$ is a vector bundle of finite rank. The differential ds gives us a homomorphism

$$T_Y|_U \xrightarrow{ds} F|_U. \qquad\qquad (27)$$

By the implicit function theorem again, if ds is surjective over U, U is smooth of dimension

$$vd(U) = \dim Y - \mathrm{rank}\, F = \mathrm{rank}\,\ker(ds|_U) - \mathrm{rank}\,\mathrm{coker}(ds|_U) \qquad (28)$$

and the tangent bundle T_U of U is the kernel of (27). As the vanishing of the cokernel of (27) implies the smoothness of U, we call the cokernel of (27) the *obstruction sheaf*[46] of the local model (26) and (27) is called a *tangent-obstruction complex*. We call (28) the *expected dimension* or the *virtual dimension* of U defined by the local model (26).

---

[44] The fiber of a vector bundle $\pi : E \to M$ over $p \in M$ is the vector space $\pi^{-1}(p)$.

[45] If M is a complex manifold, the tangent bundle T_M is a holomorphic vector bundle obviously. The complex structure on T_M is the multiplication $\sqrt{-1} : T_M \to T_M$ by $\sqrt{-1}$.

[46] If one tries to extend a map $C_m \to U$ in the mth jet of U to the $m + 1$st jet, there is an element in the obstruction sheaf whose vanishing is equivalent to the existence of an extension.

Most moduli spaces in algebraic geometry are very singular and for a finite dimensional local model (26), U usually has a bigger dimension than the expected (28). The singularity of $U = s^{-1}(0)$ arises if the section s is not transversal to the zero section 0 of F. If we perturb s slightly to obtain a section s' transversal to the zero section 0, we obtain a smooth manifold $U' = s'^{-1}(0)$ of expected dimension (28). By construction, U' is very close to U with a proper[47] map $U' \to U$. We wish to use U' instead of U to define enumerative invariants but U' depends on the choice of perturbation s'. To remove the dependency, we use a homology theory[48] and push down U' to U homologically. Algebraic topology neatly summarizes this method of perturbation of the section in terms of (refined) Euler class

$$s^! = e(F, s) : H_*(Y) \longrightarrow H_{*-2r}(U)$$

for a complex vector bundle F of rank r and a section s of F. In case Y is a complex manifold, we obtain the *virtual fundamental class* by applying $s^! = e(F, s)$ to the fundamental class[49] $[Y]$ of Y, i.e.

$$[U]^{\mathrm{vir}} := e(F, s) \cap [Y] = s^![Y] \in H_{vd_{\mathbb{R}}(U)}(U).$$

In this way, moduli spaces M we are interested in usually admit an open cover $\{M_\alpha \to M\}$ together with finite dimensional local models

$$
\begin{array}{cc}
F_\alpha & \hspace{4cm} (29) \\
\Big\uparrow \Big\downarrow {\scriptstyle s_\alpha} & \\
M_\alpha = s_\alpha^{-1}(0) \longrightarrow Y_\alpha &
\end{array}
$$

where Y_α is a complex manifold and F_α is a holomorphic vector bundle while s_α is a holomorphic section. So we have local virtual fundamental classes

$$[M_\alpha]^{\mathrm{vir}} = s_\alpha^![Y_\alpha] \in H_*(M_\alpha).$$

The question is then:

Can we glue $[M_\alpha]^{\mathrm{vir}} \in H_*(M_\alpha)$ to a homology class $[M]^{\mathrm{vir}} \in H_*(M)$?

In other words, we want a homology class $[M]^{\mathrm{vir}}$ whose restriction to M_α is $[M_\alpha]^{\mathrm{vir}}$. Of course, we cannot expect a positive answer for free. For instance, the virtual dimensions should be independent of α.

---

[47] A proper map is a continuous map such that the inverse image of a compact set is compact.

[48] In a sense, a homology theory is a way to identify submanifolds which does not change an integral of a closed differential form, by Stokes' theorem.

[49] The fundamental class of a complex manifold is the whole space together with the canonical orientation $\sqrt{-1}^n dz_1 \wedge d\bar{z}_1 \wedge \cdots \wedge dz_n \wedge d\bar{z}_n$ in local coordinates.

Let I_α be the image of the dual $s_\alpha^\vee : F_\alpha^\vee \to \mathcal{O}_{Y_\alpha}$ of the section s_α.[50] Holomorphic functions belonging to I_α determine M_α as their common zero locus. Hence the dual of the tangent-obstruction complex

$$T_{Y_\alpha}|_{M_\alpha} \xrightarrow{ds_\alpha} F_\alpha|_{M_\alpha} \tag{30}$$

fits into the commutative diagram

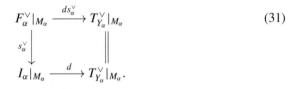

$$\tag{31}$$

For all α, the lower arrows are canonical (up to quasi-isomorphism) and always glue to a global object, called the truncated cotangent complex $\mathbb{L}_M$ of M. The correct condition discovered in 1995 by Li-Tian and Behrend-Fantechi for gluing the local virtual fundamental classes $[M_\alpha]^{\mathrm{vir}}$ to a global virtual fundamental class $[M]^{\mathrm{vir}}$ is that the tangent-obstruction complexes (30) glue to a complex $E = [E_0 \to E_1]$ of vector bundles on M[51] and (31) also glue to a morphism $\phi : E^\vee \to \mathbb{L}_M$. This complex E together with the morphism ϕ is called a *perfect obstruction theory* on M.

Given a perfect obstruction theory $E = [E_0 \to E_1]$, the normal cones C_{M_α/Y_α} induce cones

$$C_{M_\alpha/Y_\alpha} \times E_0|_{M_\alpha} / T_{Y_\alpha}|_{M_\alpha} \subset E_1|_{M_\alpha}$$

which glue to a cone $C \subset E_1$ over M. Now the *virtual fundamental class* of M is defined to be the intersection of C with a perturbation of the zero section of E_1 by

$$[M]^{\mathrm{vir}} = 0_{E_1}^![C] = e(E_1|_C, \tau) \cap [C] \tag{32}$$

where τ is the tautological section $c \mapsto (c, c)$ of

$$E_1|_C = \{(v, c) \in E_1 \times C \mid \pi(v) = \pi(c)\} \to C, \quad (v, c) \mapsto c$$

and $\pi : E_1 \to M$ is the bundle projection.[52]

---

[50] A holomorphic section $s : M \to E$ of a holomorphic vector bundle $\pi : E \to M$ can be thought of as a map $\mathcal{O}_M \to E$ of the trivial bundle $\mathcal{O}_M = M \times \mathbb{C}$ into E which maps (x, λ) to $\lambda s(x)$ for $x \in M$.

[51] More precisely, we only need a derived category object $E \in D^b(\mathrm{Coh}\, M)$ whose restriction to M_α is isomorphic to (30). Li-Tian's condition is weaker than this, now called a *semi-perfect obstruction theory*.

[52] Nowadays, the construction of the virtual fundamental class $[M]^{\mathrm{vir}}$ requires much less assumptions: We don't need a global resolution $[E_0 \to E_1]$ by vector bundles and we only need a derived category object which locally is isomorphic to (30). In fact, we don't even need a global derived category object, just local objects (30) together with a compatibility condition on the obstruction

Since 1995, lots of enumerative invariants have been defined by integrals of cohomology classes against virtual fundamental classes. Enumerative invariants defined in this way are sometimes called *virtual invariants* and have been the focus of intensive research during the past 25 years in enumerative geometry because they automatically satisfy nice properties like deformation invariance under reasonable assumptions.

Unfortunately, it is not easy to handle the virtual fundamental classes and usually computation of virtual invariants is extremely difficult. As far as I know, there are only three techniques to handle virtual fundamental classes as follows.

1. Lefschetz hyperplane principle: If $\iota : M \hookrightarrow N$ is a closed immersion and the normal cone $C_{M/N}$ embeds into $V|_M$ for some holomorphic vector bundle V on N,[53] we can compare the virtual fundamental classes of M and N by

$$[M]^{\mathrm{vir}} = \iota^![N]^{\mathrm{vir}}, \quad \iota_*[M]^{\mathrm{vir}} = e(V) \cap [N]^{\mathrm{vir}} \qquad (33)$$

   where $e(V)$ denotes the Euler class of V.[54]

2. Torus localization: Let M be equipped with an action of the multiplicative group $\mathbb{C}^*$ and a $\mathbb{C}^*$-equivariant perfect obstruction theory $E = [E_0 \to E_1]$. Then the virtual fundamental class is localized to the fixed point locus $M^{\mathbb{C}^*}$ by the formula

$$[M]^{\mathrm{vir}} = \frac{[M^{\mathbb{C}^*}]^{\mathrm{vir}}}{e(N^{\mathrm{vir}})}, \quad \text{where } e(N^{\mathrm{vir}}) = \frac{e(E_0|_{M^{\mathbb{C}^*}}^{mv})}{e(E_1|_{M^{\mathbb{C}^*}}^{mv})} \qquad (34)$$

   and $E_i|_{M^{\mathbb{C}^*}}^{mv}$ $(i = 0, 1)$ denotes the subbundle of $E_i|_{M^{\mathbb{C}^*}}$ where $\mathbb{C}^*$ acts with nontrivial weights.

3. Cosection localization: Let M be equipped with a perfect obstruction theory $[E_0 \to E_1]$ whose cokernel Ob_M admits a homomorphism $\sigma : Ob_M \to \mathcal{O}_M$ to the trivial bundle. We call σ a *cosection* of Ob_M. Then $[M]^{\mathrm{vir}}$ is localized to a class $[M]^{\mathrm{vir}}_{\mathrm{loc}}$ supported in the zero locus of σ. Often $\sigma^{-1}(0)$ is much simpler than M, sometimes just the empty set or a smooth point, and hence the cosection localization gives us a vanishing result or simplifies the computation of a virtual invariant.

---

sheaf and obstruction assignments. However, for convenience of explanation, I will assume that we have a global resolution $[E_0 \to E_1]$ of the perfect obstruction theory.

[53] We further need that the perfect obstruction theories E_M and E_N for M and N fit into a commutative diagram of exact triangles

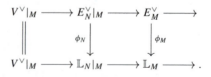

[54] There is a more general form of the (quantum) Lefschetz hyperplane principle, namely the functoriality of *virtual pullbacks*.

Often a combination of these techniques turns out to be quite powerful. For instance, one can compute the genus 0 GW invariants of a hypersurface in a projective space by first pushing the computation to the projective space by the Lefschetz hyperplane principle and then localizing to the fixed point locus by the torus localization. One way to use the torus or cosection localization is to define a virtual invariant even when the moduli space M is not compact. In general, we cannot integrate a cohomology class over a noncompact space. If there is a $\mathbb{C}^*$ action on M and the fixed point set $M^{\mathbb{C}^*}$ is compact, we can define $[M]^{\mathrm{vir}}$ by (34). If there is a cosection σ of Ob_M and $\sigma^{-1}(0)$ is compact, the cosection localized virtual cycle $[M]^{\mathrm{vir}}_{\mathrm{loc}}$ is compactly supported and hence we can integrate cohomology classes against $[M]^{\mathrm{vir}}_{\mathrm{loc}}$.

4.3   Step 3: Virtual Invariants

As we mentioned before, virtual invariants are integrals of cohomology classes against virtual fundamental classes. For the moduli space $\overline{M}_{g,n}(Q, d)$ of stable maps to a smooth projective variety Q in Sect. 4.1, there is a natural perfect obstruction theory.[55] By the recipe of Sect. 4.2, we thus obtain the virtual fundamental class

$$[\overline{M}_{g,n}(Q, d)]^{\mathrm{vir}} \in H_*(\overline{M}_{g,n}(Q, d))$$

and the *Gromov-Witten* (GW for short) invariants of Q are defined as integrals of cohomology classes against $[\overline{M}_{g,n}(Q, d)]^{\mathrm{vir}}$. In particular, when Q is a Calabi-Yau 3-fold, the expected dimension of $\overline{M}_{g,0}(Q, d)$ is zero for any $d \in H_2(Q, \mathbb{Z})$ and thus $[\overline{M}_{g,0}(Q, d)]^{\mathrm{vir}} \in H_0(\overline{M}_{g,0}(Q, d))$, a linear combination of points. By taking the sum of the coefficients, we obtain the GW invariant

$$GW^Q_{g,d} = \deg [\overline{M}_{g,0}(Q, d)]^{\mathrm{vir}} \in \mathbb{Q}. \tag{35}$$

When Q is a general quintic hypersurface in $\mathbb{P}^4$, there are only finitely many rational curves of degree $d \le 7$ and it is conjectured that the same should be true for all d. Hence the count of rational curves of degree d in Q (Question 10) can be taken care of by the GW invariant of Q if we remove contributions from multiple covers.

Let us sketch how Givental and Lian-Liu-Yau computed the genus 0 Gromov-Witten invariant of a quintic 3-fold Q in $\mathbb{P}^4$. By composing any stable map to Q with the inclusion $Q \subset \mathbb{P}^4$, we have an inclusion

$$X := \overline{M}_{0,0}(Q, d) \subset \overline{M}_{0,0}(\mathbb{P}^4, d) =: Y$$

into the smooth projective variety Y of dimension $5d + 1$. The universal family

---

[55] The deformation theory of a holomorphic map $f : C \to Q$ was well studied since 1960s. The perfect obstruction theory at a point $(f : C \to Q) \in \overline{M}_{g,n}(Q, d)$ comes from the hyperext groups $\mathrm{Ext}^*([f^*\Omega_Q \to \Omega_C^{\log}], \mathcal{O}_C)$ where $\Omega_Q = T_Q^\vee$ denotes the cotangent bundle of Q.

of stable maps gives us a vector bundle $N = \pi_* f^* \mathcal{O}_{\mathbb{P}^4}(5)$ of rank $5d + 1$ and the defining quintic polynomial of Q pulls back to a section s of N by the homomorphism

$$H^0(\mathbb{P}^4, \mathcal{O}_{\mathbb{P}^4}(5)) \xrightarrow{f^*} H^0(\mathcal{C}, f^* \mathcal{O}_{\mathbb{P}^4}(5)) = H^0(Y, N)$$

so that we have a diagram

$$
\begin{array}{c}
N = \pi_* f^* \mathcal{O}_{\mathbb{P}^4}(5) \ . \\
s \nearrow \Big\uparrow \Big\downarrow \\
X = s^{-1}(0) \overset{\iota}{\hookrightarrow} Y
\end{array}
$$

By the Lefschetz hyperplane principle, we have

$$[X]^{\mathrm{vir}} = e(N, s) \cap [Y] \in H_0(X).$$

The genus 0 GW invariant of Q is thus

$$\int_{[X]^{\mathrm{vir}}} 1 = \int_Y e(N) \in \mathbb{Q}. \tag{36}$$

So the computation of the GW invariant of Q was pushed to Y but still Y is not easy for an explicit computation. We further push the computation to the projective space $W_d = \mathbb{P}^{5d+4}$ by the diagram

$$
\begin{array}{ccc}
\overline{M}_{0,0}(\mathbb{P}^1 \times \mathbb{P}^4, (1, d)) & \longrightarrow & \mathbb{P}(H^0(\mathbb{P}^1, \mathcal{O}(d)) \otimes \mathbb{C}^5) = W_d \\
\downarrow & & \\
X = \overline{M}_{0,0}(Q, d) \hookrightarrow Y = \overline{M}_{0,0}(\mathbb{P}^4, d) & &
\end{array}
$$

where the vertical arrow is the composition of stable maps $f : C \to \mathbb{P}^1 \times \mathbb{P}^4$ with the projection $\mathbb{P}^1 \times \mathbb{P}^4 \to \mathbb{P}^4$ and the top horizontal arrow is obtained by allowing the linear system to have base points and contracting components of the domain C whose degrees are 0 in the $\mathbb{P}^1$ direction. Except for X, the action of $(\mathbb{C}^*)^5$ by multiplication on homogeneous coordinates of $\mathbb{P}^4$ makes the diagram $(\mathbb{C}^*)^5$-equivariant. By establishing a functorial torus localization, one can push the computation of (36) to the projective space W_d where integrals of cohomology classes are straightforward.

In this way, the generating function of genus 0 GW invariants (35) is given by a hypergeometric series.

A general element of W_d represents a base point free linear system on $\mathbb{P}^1$ of degree d and thus defines a map $\mathbb{P}^1 \to \mathbb{P}^4$. A *quasi-map* from a curve C to $\mathbb{P}^r$ is defined as a line bundle L on C together with $r + 1$ sections $s_0, \cdots, s_r$ of L. In particular, W_d is the moduli space of nontrivial quasi-maps from $\mathbb{P}^1$ to $\mathbb{P}^4$ of degree d.

4.4 *More Virtual Invariants*

We've seen the three steps to define a virtual invariant. So if you can find a compact[56] moduli space M equipped with a perfect obstruction theory, you have a virtual invariant. Nowadays, there are many virtual invariants defined in this way. Let us take a look at a few of them related to curve counting.

So we want a compactified moduli space of curves in a smooth projective variety Q which admits a perfect obstruction theory. The moduli space $\overline{M}_{g,n}(Q, d)$ is such an example. Are there any other? Let's think about the unit circle S^1 in plane centered at the origin. Probably the first definition of the unit circle S^1 you learned says that it is the set of points in plane with distance 1 from the origin. A little later, you probably learned that the unit circle is the same thing as the locus of points in the coordinate plane whose coordinates satisfy $x^2 + y^2 = 1$. A projective variety $Z \subset \mathbb{P}^n$ is uniquely determined by the homogeneous ideal $I = I(Z)$ of polynomials vanishing on Z. Recall that an ideal I of a polynomial ring $\mathbb{C}[z_0, \cdots, z_n]$ is homogeneous if it is generated by homogeneous polynomials. We may write $I = \oplus_{d \geq 0} I_d$ where $I_d = I \cap \mathbb{C}[z_0, \cdots, z_n]_d$ and $\mathbb{C}[z_0, \cdots, z_n]_d$ is the vector space spanned by homogeneous polynomials in $z_0, \cdots, z_n$ of degree d. Two homogeneous (radical) ideals I and I' in $\mathbb{C}[z_0, \cdots, z_n]$ define the same projective subvariety in $\mathbb{P}^n$ if and only if $I_d = I'_d$ for $d \gg 0$ and we write $I \sim I'$. So you may think of a projective curve in $\mathbb{P}^n$ as a homogeneous ideal I up to the equivalence $\sim$. For any homogeneous ideal I of $\mathbb{C}[z_0, \cdots, z_n]$, $\dim \mathbb{C}[z_0, \cdots, z_n]_d / I_d$ is a polynomial function of d for $d \gg 0$ whose defining polynomial is called the *Hilbert polynomial*. Grothendieck proved that the set of homogeneous ideals $I \subset \mathbb{C}[z_0, \cdots, z_n]$ up to $\sim$ with a fixed Hilbert polynomial forms a projective scheme,[57] called the *Hilbert scheme*. From this it is straightforward to conclude that the set of all projective subschemes with a fixed Hilbert polynomial in a projective variety Q is a projective scheme M in a natural way.[58]

---

[56] If the moduli space is not compact, you can still define a virtual invariant by finding a torus action whose fixed locus is compact or by finding a cosection of the obstruction sheaf whose zero locus is compact.

[57] A projective scheme is a generalization of a projective variety, in that projective varieties in $\mathbb{P}^n$ are in bijection with radical homogeneous ideals in $\mathbb{C}[z_0, \cdots, z_n]$ up to $\sim$ while projective schemes are in bijection with all homogeneous ideals strictly contained in the maximal ideal $(z_0, z_1, \cdots, z_n)$.

[58] A homogeneous ideal $I_C \subset \mathbb{C}[z_0, \cdots, z_n]$ contains the ideal I_Q of Q if and only if the composition $I_Q \hookrightarrow \mathbb{C}[z_0, \cdots, z_n] \to \mathbb{C}[z_0, \cdots, z_n]/I_C$ is zero.

Let $M = M_P(Q)$ be the Hilbert scheme of curves (more generally projective subschemes) with fixed Hilbert polynomial P in a Calabi-Yau 3-fold Q. Then M admits a perfect obstruction theory[59] with expected dimension 0. We thus obtain the virtual fundamental class $[M]^{vir} \in H_0(M)$ whose degree[60]

$$DT_P^Q = \deg [M_P(Q)]^{vir} \tag{37}$$

is defined to be the *Donaldson-Thomas* (DT for short) invariant of Q. Usually DT is more difficult to compute than GW. However there are many advantages of DT over GW:

1. DT can enumerate not just curves but also projective subschemes or vector bundles.
2. The perfect obstruction theory of DT is symmetric and the tangent space is dual to the obstruction space at each point.
3. The DT moduli space is locally the critical locus of a holomorphic function on a complex manifold.
4. DT is an integer while GW is a rational number.

As GW and DT both can enumerate curves in a smooth projective variety Q, they should be related somehow. The Maulik-Nekrasov-Okounkov-Pandharipande conjecture states a precise formula for the equivalence of GW and DT after a suitable change of coordinates and removing contributions by points floating away from curves etc. It was proved for quintic 3-folds and some other cases.

There are still many more ways to think of a curve C in a smooth projective variety Q. We can think of the coordinate ring $\mathbb{C}[z_0, \cdots, z_n]/I_C$ of C as a homogeneous module over $\mathbb{C}[z_0, \cdots, z_n]/I_Q$.[61] This perspective gives us another way to define a virtual invariant enumerating curves in Q, also called the DT invariant. We can think of a curve C in Q as a homomorphism

$$\mathcal{O}_Q = \mathbb{C}[z_0, \cdots, z_n]/I_Q \to \mathbb{C}[z_0, \cdots, z_n]/I_C = \mathcal{O}_C$$

and this perspective gives us the virtual invariant of stable pairs, called the Pandharipande-Thomas (PT for short) invariant. We can also think of a curve as a quasi-map and hence construct a virtual invariant.

Any reasonably defined curve counting virtual invariant should be related with known invariants like GW, DT and PT.

---

[59] The tangent and obstruction spaces at $I \in M$ are the traceless ext-groups $Ext_Q^i(I, I)_0$ for $i = 1, 2$.

[60] $\deg : H_0(M) \to H_0(pt) \cong \mathbb{Q}$ is the pushforward by the proper map $M \to pt$. This map is counting points with multiplicity.

[61] So we can think of a curve C in Q as a coherent sheaf $\mathcal{O}_C$ on Q. By GIT, we know that we have to confine ourselves to the study of *semistable* sheaves and there is a projective stack M parameterizing semistable sheaves on Q after fixing an ample line bundle. The open set of stable sheaves admits a perfect obstruction theory (when $\dim Q = 3$ and $K_Q \leq 0$) and hence we obtain a DT invariant when all semistable sheaves are stable. If there are semistable sheaves which is not stable, we need to use the Hall algebra machinery or the Kirwan blowup, to obtain a generalized DT invariant.

5   Recent Developments

To end this note, let us review some of recent developments in enumerative geometry. Inevitably, I have to be very sketchy and cannot explain all the terms and ideas, because each of the topics deserves an independent survey article.

5.1   Cohomological, K-Theoretic and Motivic Refinements and Wall Crossing

For a smooth projective variety Q, the collection of all vector bundles over Q with fixed topological type (Chern classes) is usually not compact and we need to add coherent sheaves to compactify it. Roughly speaking, a coherent sheaf is like a vector bundle on a subvariety with rounding off along the boundaries. The collection of all coherent sheaves is big enough for compactification but it is not Hausdorff. So we need to take out some undesired coherent sheaves and a criterion to determine which sheaves to delete is called a stability condition. A most popular choice - Gieseker stability - comes from geometric invariant theory and it is given by an inequality of Hilbert polynomials. After fixing a stability condition, under favorable circumstances, the collection of all stable coherent sheaves on Q forms a compact Hausdorff space which admits a virtual fundamental class when Q is a Calabi-Yau 3-fold. So, the DT invariants of Calabi-Yau 3-folds depend on the choice of a stability condition. A wall crossing formula records the differences of DT invariants as we vary the stability condition.

If we choose the stability condition in a clever way, the DT invariants are easy to compute and hence the DT invariants for arbitrary stability conditions can be computed if we know all the wall crossings. Of course, wall crossing is really useful and has been a key tool for at least three decades in algebraic geometry.

The DT invariants are known to be motivic which means that we can cut the moduli space M into locally closed pieces and add the (weighted) DT invariants of the pieces to get the DT invariant associated to M. This motivic property enables us to employ the Hall algebra machinery and the wall crossing can be handled in a completely systematic way. Furthermore, this Hall algebra technique enables us to define DT invariants even when the circumstances are not favorable.[62]

When Q is a Calabi-Yau 3-fold, the moduli space of stable sheaves on Q is locally the critical locus of a regular function on a smooth variety. Using the motivic vanishing cycles, Kontsevich and Soibelman generalized the DT invariant to a motivic invariant which takes values in a motivic ring and studied their wall crossing.

Another refinement of DT is by K-theory. The virtual fundamental class may be defined as a class in the K-group of coherent sheaves by taking the K-theoretic intersection of the cone C with the zero section of E_1 in the notation of (32). By

---

[62] Like when there are semistable sheaves which are not stable.

taking the holomorphic Euler characteristic of a vector bundle twisted by the K-theoretic virtual fundamental class, we obtain the K-theoretic DT invariant. As K-theory is more refined than homology by (virtual) Riemann-Roch, the K-theoretic DT invariants contain more information about the geometry of Q. Recently, interesting links of K-theoretic DT with representation theory have been studied by Okounkov and others.

Finally, there is a cohomological refinement of DT invariants which may produce a correct mathematical theory of the Gopakumar-Vafa invariant or BPS invariant. The critical locus of a regular function on a smooth variety comes with a perverse sheaf of vanishing cycles. One can glue the locally defined perverse sheaves of vanishing cycles to a globally defined perverse sheaf P when the moduli space M is oriented.[63] So we can consider the hypercohomology $\mathbf{H}^*(M, P)$ of P whose Euler characteristic is the usual DT invariant. Now if the perverse sheaf P is semisimple, as it underlies a mixed Hodge module, the morphism to Chow scheme defines an action of $sl_2 \times sl_2$ on $\mathbf{H}^*(M, P)$ by the hard Lefschetz property. The first sl_2 action comes from the hard Lefschetz on the cohomology of fibers and the second sl_2 action comes from the base.

Imagine a coherent sheaf is a line bundle on a subvariety and M is the space of pairs (C, L) where $C \subset Q$ is a smooth curve and L is a line bundle on C. The morphism to Chow is the forgetful map $(C, L) \mapsto C$ whose fiber over C is the Jacobian of C, an abelian variety, which is topologically a product J_g of g copies ellipses where g is the genus of C. Gopakumar-Vafa's recipe to enumerate curves of genus g in Q is to write

$$\mathbf{H}^*(M, P) \cong \bigoplus_g H^*(J_g) \otimes R_g$$

by using the $sl_2 \times sl_2$ action and then calculate the Euler characteristic of R_g. The conjecture is that the invariants defined in this way contain equivalent information as GW invariants. GV's invariants are integers and supposed to be free from multiple cover contributions - more purified curve counting invariants.

5.2 Quantum Singularity Theory

Quantum singularity theory is an attempt to enumerate curves in a smooth projective variety by the singularity of the associated affine cone. Let me try to explain the ideas for the case of Fermat quintic 3-fold

$$Q = (\sum_{i=1}^{5} x_i^5 = 0) \subset \mathbb{P}^4.$$

The general case is not more difficult.

---

[63] A moduli space of stable coherent sheaves is always orientable when there is a universal family.

Consider the quadruple (C, L, x, p) of

1. a projective curve C with at worst nodal singularities,
2. an algebraic line bundle L on C,
3. five sections $x = (x_1, \cdots, x_5) \in H^0(L)^{\oplus 5}$ of L, and
4. a section $p \in H^0(L^{-5}\omega_C)$ where ω_C is the dualizing line bundle (the sheaf of 1-forms).

The collection of all such quadruples (C, L, x, p) forms an *algebraic stack*, as it is locally the quotient of an affine scheme by a group action. Let us denote the algebraic stack by $\mathfrak{X}$.

If $x : \mathcal{O}_C^{\oplus 5} \to L$ is surjective, it induces a morphism $f : C \to \mathbb{P}^4$. Let us denote the open substack of quadruples with $x : \mathcal{O}_C^{\oplus 5} \to L$ surjective by $\mathfrak{X}_+$. The moduli space

$$\overline{M}_{g,n}(\mathbb{P}^4, d)^p = \{(C, L, x, p) \mid (C, L, x) \in \overline{M}_{g,n}(\mathbb{P}^4, d),\ p \in H^0(L^{-5}\omega_C)\}/\cong$$

of stable maps to $\mathbb{P}^4$ with p-fields is an open substack in $\mathfrak{X}_+$.

If $p : L^5 \to \omega_C$ is surjective, it has to be an isomorphism and we say (C, L) is a 5-spin curve. Let us denote the open substack of quadruples with $p : L^5 \to \omega_C$ surjective by $\mathfrak{X}_-$. The moduli space

$$S_{g,n} = \{(C, L, p) \mid p : L^5 \cong \omega_C,\ C \in \overline{M}_{g,n}\}/\cong$$

of 5-spin curves is a compact space and hence integrals of cohomology classes are well defined. Obviously $S_{g,n}$ is closed in $\mathfrak{X}_-$ and the space

$$W_{g,n} = \{(C, L, x, p) \mid (C, L, p) \in S_{g,n},\ x \in H^0(L)^{\oplus 5}\}/\cong$$

is open in $\mathfrak{X}_-$.

Here are the key points of the quantum singularity theory for $\sum_{i=1}^5 x_i^5$:

1. Any open separated Deligne-Mumford substack U of $\mathfrak{X}$ admits a perfect obstruction theory whose obstruction sheaf Ob_U admits a cosection σ.
2. For $\overline{M}_{g,n}(\mathbb{P}^4, d)^p$, the cosection localized virtual fundamental class has support in the moduli space $\overline{M}_{g,n}(Q, d)$ of stable maps to Q and satisfies

$$[\overline{M}_{g,n}(Q, d)]^{\mathrm{vir}} = \pm[\overline{M}_{g,n}(\mathbb{P}^4, d)^p]^{\mathrm{vir}}_{\mathrm{loc}}. \tag{38}$$

3. For $W_{g,n}$, the cosection localized virtual fundamental class has support in the moduli space $S_{g,n}$ of 5-spin curves and hence integrating against $[W_{g,n}]^{\mathrm{vir}}_{\mathrm{loc}}$ enables us to define virtual invariants, called the Fan-Jarvis-Ruan-Witten (FJRW for short) invariants.
4. There are ways to compare the virtual invariants by $[\overline{M}_{g,n}(\mathbb{P}^4, d)^p]^{\mathrm{vir}}_{\mathrm{loc}}$ (i.e. the GW invariants of Q) with the FJRW invariants.

To be precise, we should include orbifold structures on curves for the FJRW invariants and the complete set of FJRW invariants shares many nice properties with GW invariants, some of which are codified in the axioms of a Cohomological Field Theory.

The comparison of GW and FJRW turned out to be a powerful new technique to investigate the GW invariants of Calabi-Yau 3-folds and some old conjectures were proved by this method.

5.3   Donaldon-Thomas Theory for Calabi-Yau 4-Folds

An algebraic version of the Donaldson-Floer theory is as follows:

1. For each smooth projective surface S, we have the Donaldson invariant $D(S)$ which enumerates stable coherent sheaves on S.
2. For each smooth projective curve C, we should have a finite dimensional vector space $\mathcal{H}(C)$ with a pairing $* : \mathcal{H}(C) \otimes \mathcal{H}(C) \to \mathbb{Q}$ and $\mathcal{H}(\varnothing) = \mathbb{Q}$.
3. For each smooth projective surface S with a smooth divisor C, we should have a vector $DF(S, C) \in \mathcal{H}(C)$ satisfying $DF(S, \varnothing) = D(S)$.
4. If S degenerates into the gluing of two smooth projective surfaces S' and S'' along the smooth divisor C, we have the gluing formula

$$D(S) = DF(S', C) * DF(S'', C).$$

Intuitively, we are gluing a vector bundle F' on S' with a vector bundle F'' on S'' by an isomorphism $F'|_C \cong F''|_C$, to get a vector bundle on $S' \cup_C S''$ and deforming it to a vector bundle on S. So far, no one knows how to make this fantastic dream rigorous.[64]

More generally, one may dream about a theory which assigns a vector space $\mathcal{H}(S)$ to each d dimensional smooth projective variety S and a vector $D(Y, S) \in \mathcal{H}(S)$ to each $d + 1$ dimensional smooth projective variety Y which contains S as a divisor. When $S = \varnothing$, $\mathcal{H}(\varnothing) = \mathbb{Q}$ and $D(Y) = D(Y, \varnothing) \in \mathbb{Q}$ is an invariant of Y. Moreover we expect a gluing formula

$$D(Y) = D(Y', S) * D(Y'', S), \tag{39}$$

when Y degenerates into the union of smooth projective varieties Y' and Y'' glued along a smooth divisor S. When S is a Calabi-Yau 3-fold, we do have a vector space $\mathcal{H}(S) = \mathbf{H}^*(M, P)$ by the perverse sheaf P over the moduli space of stable coherent sheaves, mentioned above. So one may dream about an invariant of 4-folds which may be obtained from $\mathcal{H}(S)$ like in the algebraic Donaldson-Floer theory.

Such an invariant for a Calabi-Yau 4-fold (DT4 for short) was conceived by Donaldson and Thomas about 20 years ago but rigorous mathematical approaches

---

[64] The main difficulty is that we don't have an effective intersection theory of Artin stacks.

were developed only recently. There are two ways: differential geometric method by Cao-Leung, and the Kuranishi atlas method by Borisov-Joyce.[65] The key point is that the tangent-obstruction theory for the moduli space of stable sheaves on a Calabi-Yau 4-fold is a 3-term complex of vector bundles but we can locally write it as the sum $E \oplus E^\vee[-2]$ where E is a (real analogue of) perfect obstruction theory. After a choice of orientation on the moduli space, the local (Kuranishi) models give us the virtual fundamental class.

Here are some big open questions about DT4:

1. Does the cosection localization work for DT4? If so, we will get some vanishing results for hyperkahler 4-folds. Also it will help us in computation of DT4.
2. What is the wall crossing formula for DT4? For the usual DT invariant of a Calabi-Yau 3-fold, there is a beautiful wall crossing formula. A wall crossing for DT4 will be extremely important for computation and theory at the same time.
3. Will DT4 invariants satisfy a degeneration formula like (39) with $\mathcal{H}(S) = \mathbf{H}^*(M, P)$?

I expect that a big progress will follow from an *algebraic* construction of DT4. Differential geometry is intuitive and easier to grasp. However when proving a statement, algebraic geometry is often much more reliable.

5.4 Vafa-Witten Invariant

In mid 1990s, Vafa and Witten computed the Euler numbers of moduli spaces of semistable sheaves with fixed determinant on smooth projective surfaces, in a *physical* manner. In favorable circumstances, their Euler numbers coincide with the usual Euler numbers (up to sign). But until very recently, no one could find a mathematical theory of VW's Euler numbers in general. When the surface S is K3, VW's Euler numbers gave generating series which are modular forms!

What is the Euler number of a compact smooth manifold? By Poincaré-Hopf, if M is a compact complex manifold of dimension r, then the cotangent bundle $\Omega_M = T_M^\vee$ is a vector bundle of rank r whose Euler class $e(\Omega_M)$ gives us the Euler characteristic

$$\chi(M) = (-1)^r \int_M e(\Omega_M).$$

In particular, the number of points in the zero locus of a 1-form, if finite, is the Euler characteristic of M up to sign. How can we generalize this formula to the case where M is not smooth?

Let me try to convey VW's idea from the perspective of virtual intersection theory. As we saw in Sect. 4.2, the moduli spaces M we are interested in often admit perfect obstruction theories, which means that we have the local description (29).

---

[65] There was an announcement of a third (algebraic) approach using half-Euler class by Oh-Thomas.

Think of the open set M_α of M as the smooth manifold Y_α with constraints $s_\alpha = 0$. Do you remember the Lagrange multipliers method in calculus? If we want to deal with constraints, we should not try to cut the manifold by the constraints but instead increase the manifold by adding more variables. Geometrically, the Lagrange multipliers method says that we should embed Y_α into the dual bundle $F_\alpha^\vee$ of F_α as the zero section. This is exactly what we do in the Lagrange multipliers method: we add as many variables as the constraints and pair them by the natural pairing of F_α with $F_\alpha^\vee$. The section s_α of F_α gives us a holomorphic function

$$\hat{s}_\alpha : F_\alpha^\vee \longrightarrow \mathbb{C}, \quad \xi \mapsto \xi \cdot s_\alpha(\pi_\alpha(\xi))$$

where $\pi_\alpha : F_\alpha^\vee \to Y_\alpha$ is the bundle projection.

Suppose that $ds_\alpha : T_{Y_\alpha} \to F_\alpha$ is surjective so that M_α is smooth. Then a holomorphic function φ on Y_α which has discrete critical locus gives us a representative

$$\mathrm{Crit}(\varphi|_{M_\alpha}) = \mathrm{zero}(d\varphi|_{M_\alpha})$$

of the Euler class of the cotangent bundle Ω_{M_α}. On the other hand, $\pi_\alpha^* \varphi + \hat{s}_\alpha$ is a holomorphic function on $F_\alpha^\vee$ whose critical locus is a perturbation of $\mathrm{Crit}(\varphi|_{M_\alpha})$ by the Lagrange multipliers method. So the critical locus $\mathrm{Crit}(\pi_\alpha^* \varphi + \hat{s}_\alpha)$, a perturbation of $\mathrm{Crit}(\hat{s}_\alpha)$, on $F_\alpha^\vee$ gives us the Euler class of M_α up to sign.

Now the interesting point is that if M admits a perfect obstruction theory with local models (29), the critical loci $\{\mathrm{Crit}(\hat{s}_\alpha)\}$ glue to an algebraic stack $\mathcal{N} = Ob_M^\vee$, called the dual obstruction cone of M. It comes with local models

$$\begin{array}{ccc} & & \Omega_{F_\alpha^\vee} \\ & & \Big\downarrow\Big\uparrow {\scriptstyle d\hat{s}_\alpha} \\ \mathcal{N}_\alpha = (d\hat{s}_\alpha)^{-1}(0) & \longrightarrow & F_\alpha^\vee \end{array}$$

and hence a symmetric (semi-perfect) obstruction theory. Therefore, we can think of the virtual invariant

$$(-1)^{vd(\mathcal{N})} \int_{[\mathcal{N}]^{\mathrm{vir}}} 1$$

as the Euler number of M.

Now back to VW's Euler numbers. The moduli stack M of semistable sheaves with fixed determinant on a smooth projective surface S has the tangent and obstruction spaces

$$T_M|_E = \mathrm{Ext}_S^1(E, E)_0, \quad Ob_M|_E = \mathrm{Ext}_S^2(E, E)_0$$

at $E \in M$, where the subscript 0 denotes the traceless part. For the dual obstruction cone $\mathcal{N}$, we have to add the dual space

$$\mathrm{Hom}(E, E \otimes K_S)_0 \cong \mathrm{Ext}_S^2(E, E)_0^{\vee}$$

of the obstruction space by Serre duality where $K_S = \wedge^2 \Omega_S$ denotes the canonical line bundle of S. A homomorphism $\phi : E \to E \otimes K_S$ is called a *Higgs field* and the pair (E, ϕ) is called a *Higgs pair.*[66] Hence we should consider the moduli space $\mathcal{N}$ of Higgs pairs (E, ϕ) on S with $\det E$ fixed and $\mathrm{tr}(\phi) = 0$. Tanaka and Thomas proved that $\mathcal{N}$ admits a symmetric obstruction theory and defined the Vafa-Witten invariant as

$$\int_{[\mathcal{N}]^{\mathrm{vir}}} 1 = \int_{[\mathcal{N}^{\mathbb{C}^*}]} \frac{1}{e(N^{\mathrm{vir}})}$$

by using the torus localization, where $t \in \mathbb{C}^*$ acts on (E, ϕ) by $(E, t\phi)$. Their calculations match with the conjectured formulas by Vafa-Witten. Thomas also constructed a K-theoretic refinement of VW's Euler characteristic and computed the refined invariant in some examples.

For the case where there are strictly semistable sheaves, Tanaka-Thomas used the Mochizuki/Joyce-Song technique of stable pairs. In a sense, adding a section to a moduli problem is like blowing up the moduli space and hence this M/J-S technique should be related to the singularity at strictly semistable points. It will be interesting to find a direct way of defining the VW invariant by investigating the singularity and of establishing the modularity.

After all these years, we still do not understand the Euler characteristic so well!

The references below are never meant to be anything close to a complete list, which may easily take a dozen or more pages. The list below just indicates a few starting points for further reading.

References

1. K. Behrend, B. Fantechi, The intrinsic normal cone. Invent. Math. **128**(1), 45–88 (1997)
2. W. Fulton, *Intersection Theory*, Ergebnisse der Mathematik und ihrer Grenzgebiete. 3. Folge. A Series of Modern Surveys in Mathematics, 2, 2nd edn. (Springer, Berlin, 1998)
3. W. Fulton, R. Pandharipande, *Notes on stable maps and quantum cohomology,* Algebraic geometry, Santa Cruz 1995, 45-96, Proc. Sympos. Pure Math., 62, Part 2. (Amer. Math. Soc., Providence, RI, 1997)
4. T. Graber, R. Pandharipande, Localization of virtual classes. Invent. Math. **135**(2), 487–518 (1999)
5. Y.-H. Kiem, J. Li, Localizing virtual cycles by cosections. J. Amer. Math. Soc. **26**(4), 1025–1050 (2013)
6. F.C. Kirwan, *Complex Algebraic Curves*, London Mathematical Society Student Texts, vol. 23. (Cambridge University Press, Cambridge, 1992)

---

[66] A Higgs pair on S is the same as a coherent sheaf on the Calabi-Yau 3-fold K_S with compact support. If we don't fixed $\det E$ and $\mathrm{tr}(\phi)$, the VW invariant becomes the DT invariant of K_S which does not match VW's Euler number.

7. S.L. Kleiman, *Intersection theory and enumerative geometry: a decade in review*. Proc. Sympos. Pure Math., 46, Part 2, Algebraic geometry, Bowdoin, 1985 (Brunswick, Maine, 1985), pp. 321–370, (Amer. Math. Soc, Providence, RI, 1987)
8. J. Li, G. Tian, Virtual moduli cycles and Gromov-Witten invariants of algebraic varieties. J. Amer. Math. Soc. **11**(1), 119–174 (1998)
9. D. Mumford, *Algebraic Geometry. I. Complex Projective Varieties*. Reprint of the 1976 edition. Classics in Mathematics (Springer, Berlin, 1995)
10. D. Mumford, J. Fogarty, F. Kirwan, *Geometric Innvariant Theory*. 3rd edn. Ergebnisse der Mathematik und ihrer Grenzgebiete (2), 34 (Springer, Berlin, 1994)
11. A. Vistoli, Intersection theory on algebraic stacks and on their moduli spaces. Invent. Math. **97**(3), 613–670 (1989)

On the Singularity Problem for the Euler Equations

Dongho Chae

Abstract In this expository article we discuss the finite time singularity problem for the three dimensional incompressible Euler equations. The local in time well-posedness for the 3D Euler equations for initial data in the Sobolev space $H^k(\mathbb{R}^3)$, $k > 5/2$ is well-known. The question of the spontaneous apparition of singularity(blow-up), however, is a wide-open problem in the mathematical fluid mechanics. Here we overview some of the previous results on the problem, and present their recent updates. More specifically, after a brief review of Kato's classical local well-posedness result, we present the celebrated Beale, Kato and Majda's blow-up criterion, and its recent developments. After that, we review the results related to the Type I blow-up. Finally, we present recent studies on the singularity problem for the 2D Boussinesq equations, which is regarded as a good model problem for the axisymmetric 3D Euler equations.

1 Introduction

We consider a fluid flow with mass density $\rho = \rho(x, t)$, $(x, t) \in \mathbb{R}^3 \times [0, +\infty)$, which occupies the whole domain of $\mathbb{R}^3$. The two basic functions describing the motion of the flow are the fluid velocity $u = (u_1, u_2, u_3) = u(x, t)$ and the pressure $p = p(x, t)$ The mass conservation principle applied to any fixed domain $\Omega \subset \mathbb{R}^3$ during the fluid flows is expressed by the following equation:

$$\frac{d}{dt} \int_\Omega \rho(x, t) dx = - \int_{\partial\Omega} \rho u \cdot \nu dS, \tag{1}$$

where ν is the outward unit normal vector on $\partial\Omega$. Indeed, the left-hand side of (1) is the mass increasing rate in time for the fluid occupying Ω, while the right-hand

Partially supported by KIAS fund.

D. Chae (✉)
Department of Mathematics, Chung-Ang University, Seoul 06974, Korea
e-mail: dchae@cau.ac.kr

side of (1) represents the total mass per unit time, escaping Ω through the boundary $\partial\Omega$, and the equality of (1) is nothing but the mass conservation for fixed domain Ω. Applying the Gauss theorem to the right-hand side of (1), we find easily

$$\int_\Omega \{\rho_t + \nabla \cdot (\rho u)\}\, dx = 0,$$

which holds for any domain $\Omega \subset \mathbb{R}^3$. Therefore, we have the differential form of the mass conservation law in fluid as follows:

$$\rho_t + \nabla \cdot (\rho u) = 0. \tag{2}$$

Next, we apply the momentum balance principle, which is Newton's second law of motion, to a fluid in a ball $B(x, r) = \{y \in \mathbb{R}^3 \mid |x - y| < r\}$. Given $t \geqslant 0$, let $x(t) \in$ be the position of the fluid particle. Then, the velocity of the fluid at t satisfies $\frac{dx(t)}{dt} = u(x(t), t)$, while the acceleration is given by

$$\frac{d^2 x(t)}{dt^2} = \frac{d}{dt} u(x(t), t) = \frac{\partial u}{\partial t} + \frac{dx(t)}{dt} \cdot \nabla u(x(t), t)$$
$$= \frac{\partial u}{\partial t} + u \cdot \nabla u.$$

Therefore, the momentum of the fluid per unit volume at (x, t) is given by

$$\rho(x, t) \frac{d^2 x(t)}{dt^2} = \rho \frac{\partial u}{\partial t} + \rho u \cdot \nabla u. \tag{3}$$

The force due to the pressure on the surface $\partial B(x, r)$ is given by

$$-\int_{\partial B(x,r)} p(y, t)\nu\, dS, \tag{4}$$

where we consider only the force resulting from the normal directional contribution by the pressure. Actually in this consideration we use implicitly the assumption that the fluid is *ideal*. In the real physical situation we need to consider also the tangential part of the contribution of the pressure to the body force. Applying the Gauss theorem, the surface integral of (4) is transformed into

$$-\int_{B(x,r)} \nabla p(y, t)\, dy.$$

Hence, the force on the fluid particle at (x, t) per unit volume is given by

$$-\lim_{r \to 0} \frac{1}{|B(x, r)|} \int_{B(x,r)} \nabla p(y, t)\, dy = -\nabla p(x, t), \tag{5}$$

where we denote by $|A|$ the volume of $A \subset \mathbb{R}^3$. The momentum balance principle ensures the quality of (3) with (5), and we obtain

$$\rho \frac{\partial u}{\partial t} + \rho u \cdot \nabla u = -\nabla p. \tag{6}$$

The system (2) and (6) was derived first in 1755 by E. Euler in [38], and is called the Euler equations. For simplicity we further assume the homogeneity of the fluid, which means that $\rho(x, t) \equiv \text{constant} = 1$. In this case (2) reduces to the incompressibility condition $\nabla \cdot u = 0$, and the Euler equations become

$$(E) \begin{cases} u_t + u \cdot \nabla u = -\nabla p, \\ \nabla \cdot u = 0. \end{cases}$$

This is the homogeneous incompressible Euler equations for the ideal fluid. There are many nice textbooks and survey papers on the mathematical theories on the Euler equations [1, 3, 4, 26, 30–32, 46, 49]. In this article after brief studies of some of the basic properties of the equations, we review some of the classical results, and then survey recent progress on the singularity problems of the Euler equations.

Let us start by introducing the quantity $\omega = \nabla \times u$ called the vorticity, which has an important role in the incompressible fluid mechanics. Using the general vector calculus identity, $\nabla(u \cdot v) = u \cdot \nabla v + u \cdot \nabla v + u \times (\nabla \times v) + v \times (\nabla \times u)$, one can deduce

$$u \cdot \nabla u = -u \times (\nabla \times u) + \frac{1}{2} \nabla |u|^2.$$

Inserting this into the first equation of (E), we find a different form of the Euler equations

$$u_t - u \times \omega = -\nabla Q, \quad Q = \frac{1}{2}|u|^2 + p. \tag{7}$$

The quantity Q above is called the head pressure of the fluid. According to the Bernoulli theorem (see e.g. [29]) Q is constant along the stream lines. Taking curl of (7), and using the identity $\nabla \times (u \times \omega) = -u \cdot \nabla \omega - \omega \cdot \nabla u$, which holds for $\nabla \cdot u = 0$, we derive another form of the Euler equations, called the vorticity formulation.

$$\begin{cases} \omega_t + u \cdot \nabla \omega = \omega \cdot \nabla u, \\ \nabla \cdot u = 0, \quad \nabla \times u = \omega. \end{cases} \tag{8}$$

The second line of (8) can be viewed as a linear elliptic system for given ω. Formally, it can be solved as follows. From $\nabla \cdot u = 0$, applying the Poincaré lemma, there exists a vector field $\psi = (\psi_1, \psi_2, \psi_3)$ such that

$$u = \nabla \times \psi, \quad \nabla \cdot \psi = 0.$$

The second equation is imposed to remove extra degree of freedom, which is similar
to the gauge fixing in physics. Hence, we obtain the Poisson equation for ψ

$$\omega = \nabla \times (\nabla \times \psi) = \nabla(\nabla \cdot \psi) - \Delta\psi = -\Delta\psi. \tag{9}$$

Assuming sufficiently fast decay of ω at spatial infinity, we can solve (9), using the
Newtonian potential,

$$\psi = -\Delta^{-1}\psi = \frac{1}{4\pi} \int_{\mathbb{R}^3} \frac{\omega}{|x - y|} dy,$$

from which we obtain the Biot-Savart formula

$$u(x, t) = -\nabla \times \Delta^{-1}\omega = \frac{1}{4\pi} \int_{\mathbb{R}^3} \frac{(x - y) \times \omega(y, t)}{|x - y|^3} dy,$$

which represents the velocity in terms of the vorticity. It is also very important to
see the relation between ∇u and ω. ∇u is a matrix valued singular integral operator,
which can be computed as follows (see [49] for more details). For $h \in \mathbb{R}^3$ we have

$$
\begin{aligned}
\nabla u\, h &= -\nabla(\nabla \times \Delta^{-1}\omega)h \\
&= -PV \int_{\mathbb{R}^3} \left\{ \frac{\omega(y) \times h}{|x - y|^3} + \frac{3}{4\pi} \frac{[(x - y) \times \omega(y)] \otimes (x - y)}{|x - y|^3} h \right\} dy \\
&\quad + \frac{1}{3}\omega(x) \times h, \\
&:= PV \int_{\mathbb{R}^3} K(x - y)\omega(y)dyh + \frac{1}{3}\omega(x) \times h, \tag{10}
\end{aligned}
$$

where PV means the Cauchy principal value integral defined by

$$PV \int_{\mathbb{R}^n} K(x - y)f(y)dy = \lim_{\varepsilon \to 0} \int_{|x-y|>\varepsilon} K(x - y)f(y)dy.$$

The kernel $K(\cdot)$ in (10) is typical of the integral kernels defining singular integral
operator of the Calderon-Zygmund type, which have important roles in the harmonic
analysis (see e.g. [53]). We can therefore obtain the following closed form of the
vorticity formulation of the Euler equations

$$
\begin{cases}
\omega_t + u \cdot \nabla\omega = \omega \cdot \nabla u, \\
u(x, t) = \dfrac{1}{4\pi} \displaystyle\int_{\mathbb{R}^3} \dfrac{(x - y) \times \omega(y, t)}{|x - y|^3} dy.
\end{cases} \tag{11}
$$

Now we discuss the Lagrangian formulation of the Euler equations. Given $\alpha \in \mathbb{R}^3$
and a smooth vector field $u = u(x, t)$, let $X(\alpha, t)$ be the solution of the following

ordinary differential equations:

$$\begin{cases} \dfrac{\partial X(\alpha, t)}{\partial t} = u(X(\alpha, t), t), & t > 0 \\ X(\alpha, 0) = \alpha \end{cases} \tag{12}$$

The parametrized mapping $\alpha \mapsto X(\alpha, t)$ is called *the particle trajectory mapping* generated by $u = u(x, t)$. When u is the velocity field, which is a solution of (E), we say the associated $X(\alpha, t)$ the Lagrangian coordinate, and describing the dynamics of the fluid flows in terms of $X(\alpha, t)$ is called *the Lagrangian description.* Roughly speaking, it is a coordinate transform from a stationary observer to a moving observer following the flows. In terms of the Lagrangian coordinate one finds immediately that the evolution equation of (E) is written as

$$\frac{\partial^2 X(\alpha, t)}{\partial t^2} = -\nabla p(X(\alpha, t), t). \tag{13}$$

Another important equation associated with the Lagrangian coordinates is the following *Cauchy's formula,*

$$\omega(X(\alpha, t), t) = \omega_0(\alpha) \cdot \nabla X(\alpha, t), \tag{14}$$

where $\omega_0(\alpha) = \omega(\alpha, 0)$ is the initial vorticity. One can regard (14) as a translation of the first equation of (11) into the Lagrangian coordinates. For the details of the proof of (14) we refer [49].

Let us consider a closed curve $\mathcal{C}_0 = \{\gamma(s) \in \mathbb{R}^3 : s \in [0, 1], \gamma(0) = \gamma(1)\}$. For a solution (u, p) of (E) and the particle trajectory mapping generated by u we define

$$\mathcal{C}_t = X(\mathcal{C}_0, t) = \{X(\gamma(s), t) : s \in [0, 1], \gamma(0) = \gamma(1)\}.$$

Then, from (12) and (13) we find

$$\frac{d}{dt} \oint_{\mathcal{C}_t} u \cdot d\ell = \frac{d}{dt} \int_0^1 \frac{\partial X(\gamma(s), t)}{\partial t} \cdot \frac{\partial X(\gamma(s), t)}{\partial s} ds$$

$$= \int_0^1 \frac{\partial^2 X(\gamma(s), t), t)}{\partial t^2} \cdot \frac{\partial X(\gamma(s), t)}{\partial s} ds + \int_0^1 \frac{\partial X(\gamma(s), t), t)}{\partial t} \cdot \frac{\partial^2 X(\gamma(s), t)}{\partial t \partial s} ds$$

$$= -\int_0^1 \nabla p(X(\gamma(s), t), t) \cdot \frac{\partial X(\gamma(s), t)}{\partial s} ds + \frac{1}{2} \int_0^1 \frac{\partial}{\partial s} \left| \frac{\partial X(\gamma(s), t)}{\partial t} \right|^2 ds$$

$$= -\int_0^1 \frac{\partial}{\partial s} \left(p(X(\gamma(s), t), t) - \frac{1}{2} \left| \frac{\partial X(\gamma(s), t)}{\partial t} \right|^2 \right) ds = 0.$$

Therefore, we obtain the following Kelvin circulation theorem:

$$\oint_{C_t} u \cdot d\ell = \oint_{C_0} u \cdot d\ell \quad \forall t > 0. \tag{15}$$

Let C_0 be a vortex line of the initial vorticity $\omega_0 = \omega(\cdot, 0)$, defined by

$$C_0 = \left\{ \gamma(s) \in \mathbb{R}^3 \ : \ \frac{d}{ds} \gamma(s) = \lambda(s) \omega_0(\gamma(s)), s \in [0, 1], \right\}$$

for a real valued function function $\lambda(s) > 0$ for $s \in [0, 1]$. By reparametrization we may assume without the loss of generality that $\lambda(s) \equiv 1$. Then, we first claim $C_t = X(C_0, t) = \{X(\gamma(s), t) \ : \ s \in [0, 1]\}$ is a vortex line at $t > 0$. Indeed, from (14) we have

$$\frac{\partial X(\gamma(s), t)}{\partial s} = \frac{d\gamma(s)}{ds} \cdot \nabla X(\gamma(s), t)$$

$$= \omega_0(\gamma(s)) \cdot \nabla X(\gamma(s), t)$$

$$= \omega(X(\gamma(s), t), t), \tag{16}$$

and the claim is proved. Using (11), (13) and (16), we deduce

$$\frac{d}{dt} \int_0^1 u(X(\gamma(s), t), t) \cdot \omega(X(\gamma(s), t), t) ds$$

$$= -\int_0^1 \nabla p(X(\gamma(s), t), t) \cdot \omega(X(\gamma(s), t), t) ds$$

$$+ \int_0^1 u(X(\gamma(s), t), t) \cdot (\omega(X(\gamma(s), t), t) \cdot \nabla) u(X(\gamma(s), t), t) ds$$

$$= -\int_0^1 \frac{\partial}{\partial s} \left(p(X(\gamma(s), t), t) - \frac{1}{2} |u(X(\gamma(s), t), t)|^2 \right) ds = 0. \tag{17}$$

Therefore, we obtain the following helicity conservation along each closed vortex line

$$\int_0^1 u(X(\gamma(s), t), t) \cdot \omega(X(\gamma(s), t), t) ds = \int_0^1 u_0(\gamma(s)) \cdot \omega_0(\gamma(s)) ds \tag{18}$$

for all $t > 0$. This can be viewed as a localized version of the following helicity conservation law,

$$H = \int_{\mathbb{R}^3} u(x, t) \cdot \omega(x, t) dx = \int_{\mathbb{R}^3} u_0(x) \cdot \omega_0(x) dx \quad \forall t > 0. \tag{19}$$

The proof of (19) follows easily by taking $\frac{dH}{dt}$, and using (E), (11), and integrating by parts. The most important conservation law in the study of the Euler equation is the following energy conservation

$$E = \frac{1}{2} \int_{\mathbb{R}^3} |u(x,t)|^2 dx = \frac{1}{2} \int_{\mathbb{R}^3} |u_0|^2 dx \quad \forall t > 0. \tag{20}$$

This can be shown by multiplying (7) by u, and integrating it over $\mathbb{R}^3$, and integrating by parts. The reason why the energy is important in the mathematical fluid mechanics is that it is positive definite, while all the other conserved quantities in the 3D Euler equations have no definite signs.

2 The Local in Time Well-Posedness

In this section we briefly review the studies on the Cauchy problem of (E). For this we recall the notions of the Sobolev spaces. Let $\Omega \subset \mathbb{R}^n$ be a measurable subset of $\mathbb{R}^n$, and f be a measurable function on Ω. We denote $|A| = $ Lebesgue measure of $A \subset \mathbb{R}^n$. Let us define the L^q-norm by

$$\|f\|_{L^q} = \begin{cases} \left(\int_\Omega |f|^q dx \right)^{\frac{1}{q}}, & \text{if } 1 \leqslant q < +\infty, \\ \inf\{m : |\{x \in \mathbb{R}^n \,|\, |f(x)| > m\}| = 0\}, & \text{if } q = +\infty. \end{cases}$$

Then, the Lebesgue space for $q \in [1, \infty]$ is

$$L^q(\Omega) = \{f \text{ measureable on } \Omega : \|f\|_{L^q} < +\infty\},$$

Let $\alpha = (\alpha_1, \cdots, \alpha_n) \in (\mathbb{N} \cup \{0\})^n$ be a multi-index with $|\alpha| = \alpha_1 + \cdots + \alpha_n$. Then, the Sobolev space $W^{k,q}(\Omega)$ on $\Omega \subset \mathbb{R}^n$ for $k \in \mathbb{N}, 1 \leqslant q < +\infty$ is defined as

$$W^{k,q}(\Omega) = \left\{ f \in L^q(\Omega) : \left(\sum_{|\alpha| \leqslant k} \|D^\alpha f\|_{L^q(\Omega)}^q \right)^{\frac{1}{q}} := \|f\|_{W^{k,q}(\Omega)} < +\infty \right\},$$

where the derivative D^α is in the sense of distribution. In the case $q = 2$ we also denote $W^{k,2}(\Omega) = H^k(\Omega)$. In the case $\Omega = \mathbb{R}^n$ we have an equivalent Sobolev norm in $H^k(\mathbb{R}^n)$ defined by the Fourier transform. Let $\hat{f}$ be the Fourier transform of f defined by

$$\hat{f}(\xi) = \frac{1}{(2\pi)^{\frac{n}{2}}} \int_{\mathbb{R}^n} f(x) e^{-ix\cdot\xi} dx, \tag{21}$$

where $i = \sqrt{-1}$. The function f is recovered from $\hat{f}$ by the inverse Fourier transform defined by

$$f(x) = \frac{1}{(2\pi)^{\frac{n}{2}}} \int_{\mathbb{R}^n} \hat{f}(\xi) e^{ix\cdot\xi} dx. \tag{22}$$

Then, $f \in H^k(\mathbb{R}^n)$ if and only if

$$\left(\int_{\mathbb{R}^n} (1 + |\xi|^2)^k |\hat{f}(\xi)|^2 d\xi \right)^{\frac{1}{2}} := \|f\|_{H^k} < +\infty.$$

Using this equivalent norm, one can prove the following Sobolev inequality

$$\|f\|_{L^\infty} \leqslant C_k \|f\|_{H^k} \quad \forall k > \frac{n}{2}, \tag{23}$$

by a simple argument as follows.

Proof of (23): From (22) we find

$$|f(x)| \leqslant \int_{\mathbb{R}^n} |\hat{f}(\xi)| d\xi = \int_{\mathbb{R}^n} (1 + |\xi|^2)^{\frac{k}{2}} |\hat{f}(\xi)| (1 + |\xi|^2)^{-\frac{k}{2}} d\xi$$

$$\leqslant \left(\int_{\mathbb{R}^n} (1 + |\xi|^2)^k |\hat{f}(\xi)|^2 d\xi \right)^{\frac{1}{2}} \left(\int_{\mathbb{R}^n} \frac{1}{(1 + |\xi|^2)^k} d\xi \right)^{\frac{1}{2}}$$

$$\leqslant C_k \|f\|_{H^k}, \tag{24}$$

where we use the fact that for $k > \frac{n}{2}$ the following holds

$$\int_{\mathbb{R}^n} \frac{1}{(1 + |\xi|^2)^k} d\xi \leqslant C \int_0^\infty \frac{r^{n-1}}{(1 + r^2)^k} dr < +\infty.$$

Taking the supremum of (24) over $x \in \mathbb{R}^n$, we obtain (23). ∎

A fundamental local in time well-posedness result for the Cauchy problem of (E) is the following theorem due to Kato [40].

Theorem 2.1 *Let* $u_0 \in H^k(\mathbb{R}^3)$, $k > \frac{5}{2}$. *Then, there exists* $T = T(\|u_0\|_{H^k})$ *such that a unique solution* $u \in C([0, T); H^k(\mathbb{R}^3)) \cap AC(0, T; H^{k-1}(\mathbb{R}^3))$ *exists with* $u(\cdot, 0+) = u_0$, *where* $AC(a, b; X)$ *denotes the class of* X-*valued functions* u *such that* $t \mapsto u(t)$ *is absolutely continuous.*

(Sketch of the proof) Let $\alpha = (\alpha_1, \alpha_2, \alpha_3)$ a multi-index. We operate D^α on (E), and take $L^2(\mathbb{R}^3)$ inner product it with $D^\alpha u$. Then, summing over $|\alpha| \leqslant k$ we obtain

$$\frac{1}{2} \frac{d}{dt} \|u\|_{H^k}^2 = - \sum_{|\alpha| \leqslant k} \int_{\mathbb{R}^3} \{D^\alpha (u \cdot \nabla u) - (u \cdot \nabla) D^\alpha u\} \cdot D^\alpha u dx$$

$$- \sum_{|\alpha| \leqslant k} \int_{\mathbb{R}^3} (u \cdot \nabla) D^\alpha u \cdot D^\alpha u dx - \sum_{|\alpha| \leqslant k} \int_{\mathbb{R}^3} D^\alpha u \cdot \nabla D^\alpha p dx$$

$$:= I_1 + I_2 + I_3. \tag{25}$$

For I_3 we apply the integration by parts to have

$$I_3 = -\sum_{|\alpha|\leqslant k}\int_{\mathbb{R}^3} D^\alpha(\nabla\cdot u)D^\alpha p\,dx = 0. \tag{26}$$

Similarly, we also have for I_2,

$$I_2 = -\frac{1}{2}\sum_{|\alpha|\leqslant k}\int_{\mathbb{R}^3}(u\cdot\nabla)|D^\alpha u|^2 dx = \frac{1}{2}\sum_{|\alpha|\leqslant k}\int_{\mathbb{R}^3}(\nabla\cdot u)|D^\alpha u|^2 dx = 0. \tag{27}$$

In order to estimate I_1 we recall the following commutator estimate due to Klainerman and Majda [43]

$$\sum_{|\alpha|\leqslant k}\|D^\alpha(fg) - f D^\alpha g\|_{L^2}$$
$$\leqslant C_k\left\{\|\nabla f\|_{L^\infty}\|D^{k-1}g\|_{L^2} + \|D^k f\|_{L^2}\|g\|_{L^\infty}\right\}. \tag{28}$$

Applying (28) to I_1 with $f = D^\alpha u$, $g = \nabla u$, we obtain, using the Cauchy-Schwartz inequality

$$I_1 \leqslant \sum_{|\alpha|\leqslant k}\|D^\alpha(u\cdot\nabla u) - (u\cdot\nabla)D^\alpha u\|_{L^2}\|D^\alpha u\|_{L^2}$$
$$\leqslant C\|\nabla u\|_{L^\infty}\|u\|_{H^k}^2. \tag{29}$$

Combining (26), (27) and (29), using (23) for $k > \frac{5}{2}$, we find

$$\frac{d}{dt}\|u\|_{H^k} \leqslant C\|\nabla u\|_{L^\infty}\|u\|_{H^k} \leqslant C\|u\|_{H^k}^2. \tag{30}$$

From this differential inequality we find that

$$\|u(t)\|_{H^k} \leqslant \frac{\|u_0\|_{H^k}}{1 - C_0\|u_0\|_{H^k}t}, \tag{31}$$

and hence

$$\sup_{0<t<T}\|u(t)\|_{H^k} \leqslant 2\|u_0\|_{H^k} \quad\text{where}\quad T = \frac{1}{2C_0\|u_0\|_{H^k}}. \tag{32}$$

From (E) we have

$$\left\|\frac{\partial u}{\partial t}\right\|_{H^{k-1}} \leqslant \|u\cdot\nabla u\|_{H^{k-1}} + \|\nabla p\|_{H^{k-1}} := J_1 + J_2. \tag{33}$$

In order to estimate J_1 we recall the following product estimate of the Sobolev spaces $H^m(\mathbb{R}^n)$ (see e.g. [49]).

$$\|fg\|_{H^m} \leqslant C_m(\|f\|_{L^\infty}\|g\|_{H^m} + \|g\|_{L^\infty}\|f\|_{H^m}) \quad \forall m > \frac{n}{2}. \tag{34}$$

Applying this to J_1, we find

$$J_1 \leqslant \|u\|_{L^\infty}\|u\|_{H^k} + \|\nabla u\|_{L^\infty}\|u\|_{H^{k-1}} \leqslant C\|u\|_{H^k}^2 \leqslant C\|u_0\|_{H^k}^2, \tag{35}$$

where we use the Sobolev inequality (23) for $k > \frac{5}{2}$ and (32). In order to estimate J_2 we recall the method of estimating the pressure. Taking divergence of the first equation of (E), and using the second equation of the divergence free condition, we find

$$\Delta p = -\sum_{i,j=1}^{3} \partial_i \partial_j (u_i u_j),$$

and

$$p = -\sum_{i,j=1}^{3} \Delta^{-1}\partial_i \partial_j (u_i u_j) = \sum_{i,j=1}^{3} R_i R_j (u_i u_j), \tag{36}$$

where R_j, $j = 1, 2, 3$, is the Riesz transform on $\mathbb{R}^3$. The definition of R_j is easily understood via its Fourier transform,

$$\widehat{R_j(f)}(\xi) = i\frac{\xi_j}{|\xi|}\hat{f}(\xi),$$

where $i = \sqrt{-1}$. The following Calderon-Zygmund type estimate [53] holds for the Riesz transform

$$\|R_j f\|_{H^m} \leqslant C\|f\|_{H^m} \quad \forall m \geqslant 0. \tag{37}$$

Applying (37) to J_2, we estimate

$$J_2 \leqslant \|p\|_{H^k} \leqslant \sum_{i,j=1}^{3} \|R_i R_j (u_i u_j)\|_{H^k} \leqslant C\|u \otimes u\|_{H^k}$$

$$\leqslant C\|u\|_{L^\infty}\|u\|_{H^k} \leqslant C\|u\|_{H^k}^2 \leqslant C\|u_0\|_{H^k}^2, \tag{38}$$

where we use (34) and (32). Combining (35) and (38) with (33), we obtain

$$\left\|\frac{\partial u}{\partial t}\right\|_{H^{k-1}} \leqslant C\|u_0\|_{H^k}^2 \quad \forall t \in [0, T],$$

from which we have

$$\|u(t_2) - u(t_1)\|_{H^{k-1}} \leqslant \int_{t_1}^{t_2} \left\| \frac{\partial u(s)}{\partial s} \right\|_{H^{k-1}} ds$$

$$\leqslant C\|u_0\|_{H^k}^2 (t_2 - t_1) \quad \forall 0 < t_1 < t_2 < T.$$

Namely,

$$\|u\|_{Lip(0,T;H^{k-1})} \leqslant C\|u_0\|_{H^k}^2. \tag{39}$$

Once the a priori estimates (32) and (39) are obtained, the existence proof of a local in time solution is rather straightforward. We construct a sequence of the approximate solutions $\{u_m\}_{m\in\mathbb{N}}$ by mollification of (E) or by the Galerkin approximation. The sequence will be shown to satisfy the uniform estimate

$$\sup_{m\in\mathbb{N}} \|u_m\|_{L^\infty(0,T;H^k(\mathbb{R}^3))\cap Lip(0,T;H^{k-1}(\mathbb{R}^3))} \leqslant C\|u_0\|_{H^k}^2.$$

Applying the Lions-Aubin type compactness lemma (see e.g. [49]), there exist a subsequence u_{m_j} and the limit $u \in L^\infty(0, T; H^k(\mathbb{R}^3)) \cap Lip(0, T; H^{k-1}(\mathbb{R}^3))$ such that $u_{m_j} \to u$ in $L^\infty(0, T; H_{loc}^{k-\varepsilon}(\mathbb{R}^3))$ for all $\varepsilon > 0$. Using this strong convergence, we find that the limit $u \in L^\infty(0, T; H^k(\mathbb{R}^3)) \cap Lip(0, T; H^{k-1}(\mathbb{R}^3))$ satisfies (E). We now show the uniqueness. Let $u_1, u_2 \in L^\infty(0, T; H^k(\mathbb{R}^3))$ satisfy (E) with the pressure p_1 and p_2 and the initial data $u_{1,0}, u_{2,0} \in H^k(\mathbb{R}^3)$, respectively. Then, setting $u = u_1 - u_2$, $p = p_1 - p_2$, and subtracting the equation for u_2 from the one for u_1, we find that u satisfies

$$u_t + u_1 \cdot \nabla u + u \cdot \nabla u_2 = -\nabla p. \tag{40}$$

Taking L^2 inner product of (40) by u, and integrating by parts we obtain,

$$\frac{1}{2}\frac{d}{dt}\|u\|_{L^2}^2 = -\int_{\mathbb{R}^3} (u \cdot \nabla)u_2 \cdot u dx$$

$$\leqslant \|\nabla u_2\|_{L^\infty}\|u\|_{L^2}^2,$$

from which we deduce

$$\|u(t)\|_{L^2} \leqslant \|u_0\|_{L^2} \exp\left(\frac{1}{2}\int_0^t \|\nabla u_2(s)\|_{L^\infty} ds\right)$$

$$\leqslant \|u_0\|_{L^2} \exp\left(C\int_0^t \|u_2(s)\|_{H^k} ds\right)$$

$$\leqslant \|u_0\|_{L^2} \exp\left(CT\|u_{2,0}\|_{H^k} ds\right), \tag{41}$$

which shows that $u_1 \equiv u_2$ on $\mathbb{R}^3 \times [0, T]$ if $u_{1,0} = u_{2,0}$. ■

After the above results Kato and Ponce proved the local well-posedness in more general Sobolev spaces $W^{s,p}(\mathbb{R}^n)$, $s > \frac{n}{p} + 1$ [41]. This local well-posedness can be extended to exotic spaces such as the Besov space [12], and Triebel-Lizorkin spaces [13, 28]. Recently, the spatial decay conditions of such function class have been relaxed to allow linear growth of the velocities [20].

3 The BKM Type Blow-Up Criterion

Let $u \in C([0, T); H^k(\mathbb{R}^3)), k > \frac{5}{2}$, be a smooth solution to (E). We say that solution blows up at $t = T$ if

$$\limsup_{t \to T} \|u(t)\|_{H^k} = +\infty. \tag{42}$$

The question of finite time blow-up for (E) for a smooth initial data $u_0 \in H^k(\mathbb{R}^3)$ with $k > \frac{5}{2}$ is an outstanding open problem in the mathematical fluid mechanics. There are many survey papers [1, 4, 32], and numerical results [42, 47, 48] devoted to this problem. We also mention that in the case where the domain of the fluid has a singular boundary finite time blow-up is shown in [36]. Also in [27] authors proved apparition of singularity of (E) on the boundary of a cylinder. Our main concern here is the possibility of *interior singularity* in the whole domain for smooth initial data belonging to the above Sobolev space. In this direction one of the most celebrated results is the following theorem by Beale, Kato and Majda [2].

Theorem 3.1 (BKM criterion) *Let $u \in C([0, T); H^k(\mathbb{R}^3)), k > \frac{5}{2}$, be a local in time smooth solution to (E). Then, the solution blows at $t = T$ if and only if $\int_0^T \|\omega(t)\|_{L^\infty} dt = +\infty$.*

This theorem was later refined by Kozono and Taniuchi [45], replacing the L^∞ norm of ω by the BMO norm. See also a geometric type blow-up criterion [33, 35], controlling the blow-up in terms of the direction field $\xi = \omega/|\omega|$ of the vorticity. We recall the notion of BMO, the class of functions with bounded mean oscillations, which is first introduced by John and Nirenberg [39]. For $f \in L^1_{loc}(\mathbb{R}^n)$ let us set

$$f_{x,r} = \frac{1}{|B(x,r)|} \int_{B(x,r)} f(y) dy.$$

Then, BMO is defined by

$$BMO = \left\{ f \in L^1_{loc}(\mathbb{R}^n) : \sup_{x \in \mathbb{R}^n, r > 0} \frac{1}{|B(x,r)|} \int_{B(x,r)} |f(y) - f_{x,r}| dy =: \|f\|_{BMO} < +\infty \right\}.$$

We observe the obvious inequality, immediately from the definition

$$\|f\|_{BMO} \leqslant 2\|f\|_{L^\infty}. \tag{43}$$

It is well-known that BMO is bounded by the mapping of the singular integral operator $\mathcal{P}$ of the Calderon-Zygmund type

$$\|\mathcal{P}(f)\|_{BMO} \leqslant C\|f\|_{BMO}. \tag{44}$$

In particular we have

$$\|\nabla u\|_{BMO} \leqslant C\|\omega\|_{BMO} \tag{45}$$

(see (10)). A refined version of Theorem 3.1 due to Kozono and Taniuchi [45] is the following.

Theorem 3.2 *Let* $u \in C([0, T); H^k(\mathbb{R}^3))$, $k > \frac{5}{2}$, *be a local in time smooth solution to (E). Then, the solution blows at* $t = T$ *if and only if* $\int_0^T \|\omega(t)\|_{BMO}dt = +\infty$.

(Sketch of the proof) We recall the following version of the logarithmic Sobolev inequality in $\mathbb{R}^n$, which is the key inequality of the proof.

$$\|f\|_{L^\infty} \leqslant C(1 + \|f\|_{BMO}) \log(e + \|f\|_{H^m}) \quad \forall m > \frac{n}{2}. \tag{46}$$

Applying (45) and (46) to the first part of the estimate (30), we find

$$\frac{d}{dt}\|u\|_{H^k} \leqslant C\|\nabla u\|_{L^\infty}\|u\|_{H^k}$$
$$\leqslant C(1 + \|\nabla u\|_{BMO}) \log(e + \|u\|_{H^k})\|u\|_{H^k}$$
$$\leqslant C(1 + \|\omega\|_{BMO}) \log(e + \|u\|_{H^k})\|u\|_{H^k}.$$

Therefore, setting $a(t) = 1 + \|\omega(t)\|_{BMO}$, $y(t) = e + \|u(t)\|_{H^k}$, we obtain the differential inequality

$$\frac{dy}{dt} \leqslant Ca(t)y \log y.$$

This can be solved to lead us to

$$y(t) \leqslant y_0^{\exp(C \int_0^t a(s)ds)}.$$

Hence,

$$e + \|u(t)\|_{H^k} \leqslant (e + \|u_0\|_{H^k})^{\exp(C \int_0^t (1+\|\omega(s)\|_{BMO})ds)}.$$

This shows that

$$\limsup_{t \to T} \|u(t)\|_{H^k} = +\infty \Rightarrow \int_0^T \|\omega(t)\|_{BMO}dt = +\infty.$$

On the other hand, the following inequalities

$$\int_0^T \|\omega(t)\|_{BMO} dt \leqslant 2 \int_0^T \|\omega(t)\|_{L^\infty} dt \leqslant 2T \sup_{0<t<T} \|\nabla u(t)\|_{L^\infty} \leqslant CT \sup_{0<t<T} \|u(t)\|_{H^k},$$

where we use the Sobolev inequality (23) in the last step, show that

$$\int_0^T \|\omega(t)\|_{BMO} dt = +\infty \Rightarrow \limsup_{t \to T} \|u(t)\|_{H^k} = +\infty.$$

∎

Theorem 3.1 has been localized recently in [19]. To state the result we recall the notion of the local BMO space. For $r > 0$ and $x \in \mathbb{R}^n$ we denote $B(x, r) = \{y \in \mathbb{R}^n \mid |x - y| < r\}$, and $B(r) = B(0, r)$ below. By $BMO(B(r))$ we denote the space of all $u \in L^1(B(r))$ such that

$$|u|_{BMO(B(r))} = \sup_{\substack{z \in B(r) \\ 0<\rho\leqslant 2r}} \frac{1}{|B(z, \rho) \cap B(r)|} \int_{B(z,\rho)\cap B(r)} |u - u_{B(z,\rho)\cap B(r)}| dy < +\infty,$$

where we use the following notation for the average of u over $\Omega \subset \mathbb{R}^n$.

$$u_\Omega = \frac{1}{|\Omega|} \int_\Omega u\, dx.$$

The space $BMO(B(r))$ will be equipped with the norm

$$\|u\|_{BMO(B(r))} = |u|_{BMO(B(r))} + r^{-n} \|u\|_{L^1(B(r))}.$$

Note that $BMO(B(r))$ is continuously embedded into $L^q(B(r))$ for all $1 \leqslant q < +\infty$, and it holds

$$\|u\|_{L^q(B(r))} \leqslant cr^{\frac{n}{q}} \|u\|_{BMO(B(r))}.$$

The following is the localized version of Theorem 3.1.

Theorem 3.3 *Let $u \in C^1(B(\rho) \times (T - \rho, T))$ be a solution to (E) such that $u \in C([T - \rho, T); W^{2,q}(B(\rho))) \cap L^\infty(T - \rho, T; L^2(B(\rho)))$ for some $q \in (3, +\infty)$. If u satisfies*

$$\int_{T-\rho}^T |\omega(s)|_{BMO(B(\rho))} ds < +\infty, \tag{47}$$

then there exists no blow-up in $B(\rho) \times \{t = T\}$, namely

$$\limsup_{t \to T} \|u(t)\|_{W^{2,q}(B(r))} < +\infty \qquad \forall r \in (0, \rho).$$

(Idea of the Proof) There are three key ingredients of the proof of Theorem 3.3. The first one is the following local version of the logarithmic Sobolev inequality.

Lemma 3.4 *Let $B(r)$ be a ball in $\mathbb{R}^n$ with the radius $r > 0$. For every $u \in W^{1,q}(B(r))$, $n < q < +\infty$, the following inequality holds true*

$$\|u\|_{L^\infty(B(r))} \leqslant C(1 + \|u\|_{BMO(B(r))}) \log \left(e + c\|\nabla u\|_{L^q(B(r))} + Cr^{-1+\frac{n}{q}-\frac{n}{2}} \|u\|_{L^2(B(r))} \right)$$

(48)

with a constant $C > 0$ depending only on n and q.

The second key ingredient in the proof of Theorem 3.3 is the following localized version of the Kozono-Taniuchi inequality (see [44] for the global version).

Lemma 3.5 *Let $f, g \in BMO(B(r)) \cap L^q(B(r))$, $1 < q < +\infty$. Then $f \cdot g \in L^q(B(r))$ and it holds*

$$\|f \cdot g\|_{L^q(B(r))} \leqslant C \left(|f|_{BMO(B(r))} \|g\|_{L^q(B(r))} + |g|_{BMO(B(r))} \|f\|_{L^q(B(r))} \right)$$
$$+ Cr^{-\frac{3}{q}} \|f\|_{L^q(B(r))} \|g\|_{L^q(B(r))},$$

(49)

where the constant $C > 0$ depends on q only.

Using suitable sequence of cut-off functions, and using the above two lemmas one can have an iterative sequence of infinite inequalities for derivatives of the vorticity. In order to close this sequence of inequalities we establish the following Gronwall type iteration lemma.

Lemma 3.6 (Iteration lemma) *Let $a(t) \geqslant 0$ and $\beta_m : [t_0, t_1] \to \mathbb{R}$ be a sequences of bounded functions. Suppose there exists $K(t)$ such that*

$$|\beta_m(t)| < K(t)^m \quad \forall t \in [t_0, t_1), \forall m \in \mathbb{N}.$$

Suppose

$$\beta_m(t) \leqslant Cm + \int_{t_0}^{t} a(s)\beta_{m+1}(s)ds, \quad m \in \mathbb{N} \cup \{0\}.$$

Then the following inequality holds true for all $t \in [t_0, t_1]$

$$\beta_0(t) \leqslant C \int_{t_0}^{t} a(s)ds \, e^{\int_{t_0}^{t} a(s)ds}.$$

∎

We can also establish the similar continuation criterion for solutions belonging to the Hölder spaces [21]. For the precise statement of this result let us define the space $C^\alpha(\overline{\Omega}), 0 < \alpha \leqslant 1$, containing all Hölder continuous and bounded functions $f : \overline{\Omega} \to \mathbb{R}, n \in \mathbb{N}$, such that

$$[f]_{\alpha,\Omega} := \sup_{\substack{x,y\in\overline{\Omega} \\ x\neq y}} \frac{|f(x) - f(y)|}{|x - y|^\alpha} < +\infty.$$

The space $C^\alpha(\overline{\Omega})$ equipped by the norm $\|f\|_{C^\alpha(\bar{\Omega})} = \|f\|_{L^\infty(\Omega)} + [f]_{\alpha,\Omega}$ becomes a Banach space. Furthermore, by $C^{1,\alpha}(\overline{\Omega})$ we denote the space of all $f \in C^1(\overline{\Omega})$ with $\nabla f \in C^\alpha(\overline{\Omega})$.

Theorem 3.7 *Let $\Omega \subset \mathbb{R}^3$ be an open set. Let $u \in L^\infty_{loc}([0, T); C^{1,\alpha}(\Omega) \cap L^\infty(0, T; L^2(\Omega))$ be a local solution to the Euler equations. We assume that for every ball $B \subset \Omega$*

$$\int_0^T \|\omega(s)\|_{BMO(B)}ds < +\infty. \tag{50}$$

Then, $u \in L^\infty([0, T]; C^{1,\alpha}(K))$ for every compact $K \subset \Omega$.

The proof is more technical than that of Theorem 3.3.

   In all of the above theorems on the blow-up criterion basically the vorticity controls the finite time blow-up for the smooth solutions. In the followings we introduce a different type of criterion, which controls the blow-up of solutions in terms of the Hessian of the pressure. These are recent results by Chae and Constantin [14, 15].

Theorem 3.8 *Let $(u, p) \in C^1(\mathbb{R}^3 \times (0, T))$ be a solution of the Euler equation (E) with $u \in C([0, T); W^{2,q}(\mathbb{R}^3))$, for some $q > 3$. If*

$$\int_0^T \exp\left(\int_0^t \int_0^s \|D^2 p(\tau)\|_{L^\infty}d\tau ds\right) dt < +\infty, \tag{51}$$

then $\limsup_{t\to T} \|u(t)\|_{W^{2,q}} < +\infty$.

(*Sketch of the proof*) By direct computation we derive the following equation from the vorticity formulation of the Euler equations.

$$\frac{D^2\omega}{Dt^2} = -(\omega \cdot \nabla)\nabla p, \quad \text{where} \quad \frac{Df}{Dt} = \frac{\partial f}{\partial t} + u \cdot \nabla f. \tag{52}$$

Integrating twice the above along the particle trajectory, we have

$$|\omega(X(\alpha, t), t)| \leqslant |\omega_0(\alpha)| + |\omega_0(\alpha) \cdot \nabla u_0(\alpha)|t$$

$$+ \int_0^t \int_0^s |\omega(X(a, \tau), \tau)||D^2 p(X(\alpha, \tau), \tau)|d\tau ds, \tag{53}$$

where we use the fact

$$\frac{\partial}{\partial t} \omega(X(\alpha, t), t)\Big|_{t=0} = \omega(X(\alpha, t), t) \cdot \nabla u(X(\alpha, t), t)\Big|_{t=0}$$
$$= \omega_0(\alpha) \cdot \nabla u_0(\alpha).$$

Then, we establish the following Gronwall type lemma for the double integral inequality [15].

Lemma 3.9 *Let $\alpha = \alpha(t)$ be a non-decreasing function, and $\beta = \beta(t) \geqslant 0$ on $[a, b]$. Suppose $y(t) \geqslant 0$ on $[a, b]$, and satisfies*

$$y(t) \leqslant \alpha(t) + \int_a^t \int_a^s \beta(\tau) y(\tau) d\tau ds \quad \forall t \in [a, b].$$

Then, for all $t \in (a, b]$ we have

$$y(t) \leqslant \alpha(t) \exp\left(\int_a^t \int_a^s \beta(\tau) d\tau ds\right).$$

Applying this lemma, we obtain

$$|\omega(X(\alpha, t), t)| \leqslant (|\omega_0(\alpha)| + |\omega_0(\alpha) \cdot \nabla u_0(\alpha)|t) \exp\left(\int_0^t \int_0^s |D^2 p(X(\alpha, \tau), \tau)| d\tau ds\right),$$

and taking the supremum over $\alpha \in \mathbb{R}^3$, and integrating it over $[0, T]$, we find

$$\int_0^T \|\omega(t)\|_{L^\infty} dt \leqslant (\|\omega_0\|_{L^\infty} + T\|\omega_0(\alpha) \cdot \nabla u_0(\alpha)\|_{L^\infty}) \int_0^T \exp\left(\int_0^t \int_0^s \|D^2 p(\tau)\|_{L^\infty} d\tau ds\right) dt.$$

Applying the BKM criterion, we complete the proof.   ∎

The following is a localized version of the above theorem.

Theorem 3.10 *Let $(u, p) \in C^1(B(x_0, \rho) \times (T - \rho, T))$ be a solution to (E) with $u \in C([T - \rho, T); W^{2,q}(B(x_0, \rho))) \cap L^\infty(T - \rho, T; L^2(B(x_0, \rho)))$ for some $q \in (3, \infty)$. If*

$$\int_{T-\rho}^T \|u(t)\|_{L^\infty(B(x_0, \rho))} dt < +\infty \tag{54}$$

and

$$\int_{T-\rho}^T \exp\left(\int_{T-\rho}^t \int_{T-\rho}^s \|D^2 p(\tau)\|_{L^\infty(B(x_0, \rho))} d\tau ds\right) dt < +\infty, \tag{55}$$

then for all $r \in (0, \rho)$

$$\limsup_{t \to T} \|u(t)\|_{W^{2,q}(B(x_0,r))} < +\infty. \tag{56}$$

The above two theorems are refined, using new kinematic relations between various quantities in the fluid mechanics. We associate to a solution (u, p) of the Euler system (E) the $\mathbb{R}^{3\times3}$-valued functions $S = (S_{ij})$ and $P = (P_{ij})$, where

$$S_{ij} = \frac{1}{2}(\partial_i u_j + \partial_j u_i), \quad P_{ij} = \partial_i \partial_j p.$$

For the vorticity $\omega = \nabla \times u$ we define the direction vectors

$$\xi = \omega/|\omega|, \quad \zeta = S\xi/|S\xi|,$$

In the case $\omega(x, t) = 0$ we set $\alpha(x, t) = \rho(x, t) = 0$. Note that ξ is the *vorticity direction* vector, while ζ is the *vorticity stretching direction* vector. Then, we can show that the following kinematic relations hold.

Proposition 3.11 *Let (u, p) be a solution of (E), which belongs to $C^1(\mathbb{R}^3 \times (0, T))$. Then, the followings hold true on $\mathbb{R}^3 \times (0, T)$.*

$$D_t|S\omega| = -\zeta \cdot P\omega.$$

Using the above proposition, we can improve Theorem 3.7 as follows. Below we also use the notations $[f]_+ = \max\{f, 0\}$ and $[f]_- = \max\{-f, 0\}$.

Theorem 3.12 *Let $(u, p) \in C^1(\mathbb{R}^3 \times (0, T))$ be a solution of the Euler equation (E) with $u \in C([0, T); W^{2,q}(\mathbb{R}^3))$, for some $q > 3$. If*

$$\int_0^T \exp\left(\int_0^t \int_0^s \|[\zeta \cdot P\xi]_-(\tau)\|_{L^\infty} d\tau ds\right) dt < +\infty, \tag{57}$$

then $\limsup_{t \to T} \|u(t)\|_{W^{2,q}} < +\infty$.

Comparing the above theorem with Theorem 3.8, observing the pointwise inequality $|[\zeta \cdot P\xi]_-| \leqslant |P|$ the above theorem (and its localized version below) improve the result of Theorem 3.8. Furthermore, the above theorem implies that the dynamical changes of the signs of the scalar quantities $\zeta \cdot P\xi$ and $|S\xi|^2 - 2\alpha^2 - \rho$ are important in the phenomena of blow-up/regularity of the solutions to (E).

The following is a localized version of the above theorem.

Theorem 3.13 *Let $(u, p) \in C^1(B(x_0, r) \times (T - r, T))$ be a solution to (E) with $u \in C([T - r, T); W^{2,q}(B(x_0, r))) \cap L^\infty(T - r, T; L^2(B(x_0, r)))$ for some $q \in (3, \infty)$. We suppose*

$$\int_{T-r}^T \|u(t)\|_{L^\infty(B(x_0,r))} dt < +\infty,$$

and the following holds. Suppose

$$\int_{T-r}^{T} \exp\left(\int_{0}^{t}\int_{0}^{s} \|[\zeta \cdot P\xi]_{-}(\tau)\|_{L^{\infty}(B(x_0,r))} d\tau ds\right) dt < +\infty,$$

Then for all $\varepsilon \in (0, r)$ $\lim\sup_{t \to T} \|u(t)\|_{W^{2,q}(B(x_0,\varepsilon))} < +\infty$.

In the case of the Euler equations having axial symmetry there still exists the possibility of finite time blow-up. The finite time blow-up/global regularity in this case is also a wide-open question, and there are many interesting numerical results (see [47], and the references therein). Therefore, establishing a sharp blow-up criterion for this special case is also important.

Let u be an axisymmetric solution of the Euler equations if u solves (E), and can be written as

$$u = u^r(r, x_3, t)e_r + u^{\theta}(r, x_3, t)e_{\theta} + u^3(r, x_3, t)e_3,$$

where

$$e_r = (\frac{x_1}{r}, \frac{x_2}{r}, 0), \quad e_{\theta} = (\frac{x_2}{r}, \frac{-x_1}{r}, 0), \quad e_3 = (0, 0, 1), \quad r = \sqrt{x_1^2 + x_2^2}$$

are the canonical basis of the cylindrical coordinate system. The Euler equations for an axisymmetric solution turn into the following equations

$$\partial_t u^r + u^r \partial_r u^r + u^3 \partial_3 u^r = -\partial_r p + \frac{(u^{\theta})^2}{r}, \tag{58}$$

$$\partial_t u^{\theta} + u^r \partial_r u^{\theta} + u^3 \partial_3 u^{\theta} = -\frac{u^r u^{\theta}}{r}, \tag{59}$$

$$\partial_t u^3 + u^r \partial_r u^3 + u^3 \partial_3 u^3 = -\partial_3 p, \tag{60}$$

$$\partial_r (ru^r) + \partial_3 (ru^3) = 0. \tag{61}$$

Multiplying (59) by r, we see that ru^{θ} satisfies the transport equation

$$\partial_t (ru^{\theta}) + u^r \partial_r (ru^{\theta}) + u^3 \partial_3 (ru^{\theta}) = 0. \tag{62}$$

For the vorticity ω we get

$$\omega = \omega^r e_r + \omega^{\theta} e_{\theta} + \omega^3 e_3,$$

where

$$\omega^r = -\partial_3 u^{\theta}, \quad \omega^{\theta} = \partial_3 u^r - \partial_r u^3, \quad \omega^3 = \frac{u^{\theta}}{r} + \partial_r u^{\theta}.$$

Applying ∂_3 to (58) and applying ∂_r to (60), and taking the difference of the two equations, we obtain the following equation for ω^{θ}

$$\partial_t \omega^\theta + u^r \partial_r \omega^\theta + u^3 \partial_3 \omega^\theta = \frac{u^r \omega^\theta}{r} + \partial_3 \frac{(u^\theta)^2}{r}. \tag{63}$$

This leads to the equation

$$\partial_t \left(\frac{\omega^\theta}{r}\right) + u^r \partial_r \left(\frac{\omega^\theta}{r}\right) + u^3 \partial_3 \left(\frac{\omega^\theta}{r}\right) = \frac{\partial_3 (u^\theta)^2}{r^2}. \tag{64}$$

In the region off the axis we can have substantial improvement of the BKM criterion as follows [23].

Theorem 3.14 *Let $u \in C([0, T); W^{2,q}(\mathbb{R}^3)) \cap L^\infty(0, T; L^2(\mathbb{R}^3))$, $3 < q < +\infty$, be an axisymmetric solution to (E) in $\mathbb{R}^3 \times (0, T)$. If the following condition is fulfilled*

$$\int_0^T (T - t)\|\omega(t)\|_{L^\infty(B(x_*, R_0))} dt < +\infty, \tag{65}$$

for some ball $B(x_*, R_0) \subset \{x \in \mathbb{R}^3 \mid x_1^2 + x_2^2 > 0\}$, where $\omega = \nabla \times v$, then for all $0 < R < R_0$ it holds $u \in C([0, T], W^{2,q}(B(x_*, R)))$. In particular, this implies $u \in C([0, T], W^{2,q}(\mathbb{T}(x_*, R)))$. Here, $\mathbb{T}(x_*, R)$ stands for the torus generated by rotation of $B(x_*, R_0)$ around the axis, i.e.

$$\mathbb{T}(x_*, R) = \left\{ x \in \mathbb{R}^3 : \left(\sqrt{x_1^2 + x_2^2} - \rho_*\right)^2 + (x_3 - x_{3,*})^2 < R^2 \right\},$$

where $\rho_* = \sqrt{x_{1,*}^2 + x_{2,*}^2}$.

The main idea in the proof of this theorem is that the Eqs. (64) and (62) have a similar structure to the 2D Boussinesq equations (see Sect. 6 below for more concrete correspondence relations), which has a different scaling properties than the 3D Euler equations.

As an immediate consequence of the above theorem we have substantial improvement for the condition of the blow-up rate of the vorticity near the possible blow-up time as follows [23].

Theorem 3.15 *Let $u \in C([0, T); W^{2,q}(\mathbb{R}^3)) \cap L^\infty(0, T; L^2(\mathbb{R}^3))$, $3 < q < +\infty$, be an axisymmetric solution to (E) in $\mathbb{R}^3 \times (0, T)$. Suppose the following vorticity blow-up rate condition holds*

$$\sup_{t \in (0,T)} (T - t)^2 \left| \log \left(\frac{1}{T - t}\right) \right|^\alpha \|\omega(t)\|_{L^\infty(B(x_*, R_0))} < +\infty \tag{66}$$

for some $\alpha > 1$ and some ball $B(x_*, R_0) \subset \{x \in \mathbb{R}^3 : x_1^2 + x_2^2 > 0\}$. Then $u \in C([0, T]; W^{2,q}(\mathbb{T}(x_*, R))$ for all $0 < R < R_0$.

In particular, Theorem 3.14 says that there exists no singularity at $t = T$ in the off the axis region if the vorticity blow-up rate satisfies

$$\|\omega(t)\|_{L^{\infty}(\mathbb{R}^3)} = O\left(\frac{1}{(T-t)^{\gamma}}\right), \tag{67}$$

as $t \to T$ if $1 \leqslant \gamma < 2$. Due to the global BKM criterion, however, the singularity in this case should happen only on the axis. It would be interesting to compare this result with Tao's construction of a singular solution (see [54, Fig. 3, p.18]) for a *modified Euler system*, where $\gamma = 1$ and the set of singularity is a circle around the axis.

4 On the Type I Blow-Up

We observe that Euler system (E) has scaling property that if (u, p) is a solution, then for any $\lambda > 0$ and $\alpha \in \mathbb{R}$ the functions

$$u^{\lambda,\alpha}(x, t) = \lambda^{\alpha} u(\lambda x, \lambda^{\alpha+1} t), \quad p^{\lambda,\alpha}(x, t) = \lambda^{2\alpha} p(\lambda x, \lambda^{\alpha+1} t) \tag{68}$$

are also solutions with the initial data $u_0^{\lambda,\alpha}(x) = \lambda^{\alpha} u_0(\lambda x)$.

The case $\alpha = \frac{3}{2}$ is important for our analysis, since in this case the energy is scaling invariant. Indeed, by the energy conservation we have for $u^{\lambda} = u^{\lambda, \frac{3}{2}}$,

$$\|u^{\lambda}(t)\|_{L^2} = \|u(\lambda^{\frac{5}{2}} t)\|_{L^2} = \|u(t)\|_{L^2}.$$

Hereafter, we consider (E) in $\mathbb{R}^3 \times (-1, 0)$ and $t = 0$ is the possible first blow-up time.

Definition 4.1 One says that a solution u of (E) is *self-similar (SS)* with respect to $(0, 0)$ if there exists $\alpha > -1$ such that $u(x, t) = \lambda^{\alpha} u(\lambda x, \lambda^{\alpha+1} t)$ for all $\lambda > 1$.

Definition 4.2 A solution u is *discretely self-similar (DSS)* with respect to $(0, 0)$ if there exists $\alpha > -1$ such that $u(x, t) = \lambda^{\alpha} u(\lambda x, \lambda^{\alpha+1} t)$ for some $\lambda > 1$. For more specification we also say u is (λ, α)-DSS.

Definition 4.3 We say u blows up at $t = 0$ with *Type I* if

$$\limsup_{t \to 0}(-t)\|\nabla u(t)\|_{L^{\infty}} < +\infty. \tag{69}$$

If $\limsup_{t \to 0}(-t)\|\nabla u(t)\|_{L^{\infty}} = +\infty$, then we say it is of Type II.

In order to study self-similar solutions it is convenient to make self-similar transform from u on $\mathbb{R}^3 \times (-1, 0)$ to U on $\mathbb{R}^3 \times (0, +\infty)$ defined by

$$u(x, t) = \frac{1}{(-t)^{\frac{\alpha}{\alpha+1}}} U(y, s) \tag{70}$$

where

$$y = \frac{x}{(-t)^{\frac{1}{\alpha+1}}}, \quad s = -\log(-t).$$

U is called the profile. Then, (E) is transformed into equations for the profile

$$(SSE) \begin{cases} U_s + \dfrac{\alpha}{\alpha+1} U + \dfrac{1}{\alpha+1}(y \cdot \nabla)U + (U \cdot \nabla)U = -\nabla P, \\ \nabla \cdot U = 0. \end{cases}$$

Note that a SS solution of (E) is a stationary solution of (SSE), while a DSS solution of (E) is a time-periodic solution of (SSE),

$$U(y, s) = U(y, s + S_0), \quad S_0 = (\alpha + 1) \log \lambda.$$

A Type I solution of (E) is a global solution U of (SSE) with

$$\limsup_{t \to 0}(-t)\|\nabla u(t)\|_{L^\infty} = \limsup_{s \to +\infty} \|\nabla U(s)\|_{L^\infty} < +\infty.$$

SS and DSS obviously satisfy the above condition. Therefore, Type I blow-up scenario is a natural generalization of SS or DSS blow-up. There are many previous studies excluding SS or DSS blow-up (e.g. [5, 6, 9, 18]). Also, for the periodic solution of (SSE) one can show unique continuation type result [7].

In the case of DSS function having one point singularity one can have strong restriction to the spatial decay of the profile function, independent of the equations. We present it here.

Proposition 4.4 *Let be a* (λ, α)*−DSS function with* $\lambda > 1$*,* $\alpha \in \mathbb{R} \setminus \{-1\}$ *having one point singularity. Then*

(i)
$$|f(x, t)| \leqslant \frac{C}{(|x| + |t|^{\frac{1}{\alpha+1}})^\alpha} \quad \forall (x, t) \in \mathbb{R}^n \times (-\infty, 0] \setminus \{(0, 0)\},$$

where $C = C(\lambda, \alpha)$.

(ii) *Moreover, if*
$$|f(x, t)|(|x| + |t|^{\frac{1}{\alpha+1}})^\alpha = o(1),$$

as either $|x| + |t|^{\frac{1}{\alpha+1}} \to +\infty$ *or* $|x| + |t|^{\frac{1}{\alpha+1}} \to 0$*, which means*

$$\lim_{r \to \pm\infty} \sup_{e^r < |x| + |t|^{\frac{1}{\alpha+1}} < \lambda^2 e^r} |f(x, t)|(|x| + |t|^{\frac{1}{\alpha+1}})^\alpha = 0,$$

then

$$f = 0 \quad on \quad \mathbb{R}^n \times (-\infty, 0].$$

Therefore, if $f \not\equiv 0$ non trivial DSS function, then there exist $\{(x_k, t_k)\}$, $\{(\bar{x}_k, \bar{t}_k)\} \in \mathbb{R}^n \times (-\infty, 0] \setminus \{(0, 0)\}$ with $(x_k, t_k) \to +\infty$ and $(\bar{x}_k, \bar{t}_k) \to (0, 0)$ as $k \to +\infty$ such that

$$\limsup_{k \to \infty} (|x_k| + |t_k|^{\frac{1}{\alpha+1}})^\alpha |f(x_k, t_k)| > 0,$$

and

$$\limsup_{k \to 0} (|\bar{x}_k| + |\bar{t}_k|^{\frac{1}{\alpha+1}})^\alpha |f(\bar{x}_k, \bar{t}_k)| > 0.$$

(Proof) Let us define $Q_1 = B(0, \lambda) \times (-\lambda^{\alpha+1}, 0)$ and $Q_0 = B(0, 1) \times (-1, 0)$, and set $A_1 = Q_1 - Q_0$. For each $(x, t) \in \mathbb{R}^n \times (-\infty, 0] \setminus \{(0, 0)\}$ there exist an integer $k \in \mathbb{Z}$ and $(z, \tau) \in A_1$ such that $x = \lambda^k z$, $t = \lambda^{(\alpha+1)k} \tau$. Then, by the DSS property of f we have

$$(|x| + |t|^{\frac{1}{\alpha+1}})^\alpha f(x, t) = (|z| + |\tau|^{\frac{1}{\alpha+1}})^\alpha \lambda^{\alpha k} f(\lambda^k z, \lambda^{(\alpha+1)k} \tau)$$
$$= (|z| + |\tau|^{\frac{1}{\alpha+1}})^\alpha f(z, \tau). \tag{71}$$

For (i) we observe

$$|f(x, t)| = \left(\frac{|z| + |\tau|^{\frac{1}{\alpha+1}}}{|x| + |t|^{\frac{1}{\alpha+1}}} \right)^\alpha |f(z, \tau)| \leqslant \frac{C_1}{(|x| + |t|^{\frac{1}{\alpha+1}})^\alpha}$$

for all $(x, t) \in \mathbb{R}^n \times (-\infty, 0] \setminus \{(0, 0)\}$, where we set

$$C_1 = ess \sup_{(z, \tau) \in A_1} \left\{ (|z| + |\tau|^{\frac{1}{\alpha+1}})^\alpha |f(z, \tau)| \right\}.$$

In order to show (ii) we see that (71) implies also

$$\sup_{\lambda e^r < |x| + |t|^{\frac{1}{\alpha+1}} < \lambda^2 e^r} (|x| + |t|^{\frac{1}{\alpha+1}})^\alpha |f(x, t)| = \sup_{(z, \tau) \in A_1} \left\{ (|z| + |\tau|^{\frac{1}{\alpha+1}})^\alpha |f(z, \tau)| \right\},$$

from which, passing $r \to \pm\infty$, we obtain $f = 0$. $\blacksquare$

Let us consider the profile $F = F(y, s)$ of $f(x, t)$ defined by

$$f(x, t) = \frac{1}{(-t)^{\frac{\alpha}{\alpha+1}}} F(y, s) \quad with \quad y = \frac{x}{(-t)^{\frac{1}{\alpha+1}}}, \quad s = -\log(-t). \tag{72}$$

Then, by the similar argument to the above proof one can show the following:

$$\sup_{t\in(-\infty,0)} |D^m f(x,t)|(|x|+|t|^{\frac{1}{\alpha+1}})^{m+\alpha} = \sup_{s\in\mathbb{R}} |D^m F(y,s)|(|y|+1)^{m+\alpha} \quad (73)$$

for all $m \in \mathbb{N} \cup \{0\}$. Following the same argument as the above proposition we have the following.

Corollary 4.5 *Let f be a (λ, α)-DSS function, having one point singularity, and let F be its profile defined by (72). Then, there exists a constant $C > 0$ such that*

$$\sup_{s\in\mathbb{R}} |D^m F(y,s)| \leq \frac{C}{(|y|+1)^{m+\alpha}} \quad \forall y \in \mathbb{R}^n. \quad (74)$$

At this moment we could not exclude general Type I blow-up scenario for the solution of (E). As we shall observe below, however, under some smallness condition, we can remove the Type I blow-up. In this direction the following result is first derived in [8].

Theorem 4.6 *Let $u \in C([-1,0); H^m(\mathbb{R}^3))$, $m > 5/2$, be a solution to the Euler equations. Suppose u satisfies the following "small Type I condition"*

$$\limsup_{t\to 0}(-t)\|\nabla u(t)\|_{L^\infty} < 1. \quad (75)$$

Then,

$$\limsup_{t\to 0}\|u(t)\|_{H^m} < +\infty. \quad (76)$$

In other words, there exist no small Type I blow-up.

(Proof) The condition (75) implies that there exists $t_0 \in (-1,0)$ and $0 < C_0 < 1$ such that

$$\sup_{t_0 < s < 0}(-s)\|\nabla u(s)\|_{L^\infty} \leq C_0.$$

We consider the particle trajectory $X(a,t)$ generated by $u = u(x,t)$, i.e.

$$\partial_t X(a,t) = u(X(a,t),t), \qquad X(a,0) = a.$$

The vorticity form of the Euler equations

$$\partial_t \omega + u \cdot \nabla \omega = \omega \cdot \nabla u$$

can be written as an equation along the particle trajectory

$$\partial_t \{\omega(X(a,t),t)\} = (\omega \cdot \nabla u)(X(a,t),t).$$

Integrating $|\omega(X(a,t),t)|$ over $[t_0, t]$ along the particle trajectory, we obtain

$$|\omega(X(a,t),t)| \leqslant |\omega(X(a,t_0),t_0)| \exp\left(\int_{t_0}^t |\nabla u(X(a,s),s)|ds\right).$$

From this we estimate

$$\|\omega(t)\|_{L^\infty} \leqslant \|\omega(t_0)\|_{L^\infty} \exp\left(\int_{t_0}^t \|\nabla v(s)\|_{L^\infty}ds\right)$$

$$\leqslant \|\omega(t_0)\|_{L^\infty} \exp\left(C_0 \int_{t_0}^t (-s)^{-1}ds\right)$$

$$= \|\omega(t_0)\|_{L^\infty} \left(\frac{t_0}{t}\right)^{C_0} \quad \forall t \in (t_0,0).$$

Since $0 < C_0 < 1$, we have $\int_{t_0}^T \|\omega(t)\|_{L^\infty}dt < +\infty$, and by the BKM criterion above there exists no blow-up at T. ∎

The above theorem has been localized in [24] as follows.

Theorem 4.7 *Let $u \in L^\infty(-1,0; L^2(B(r))) \cap C([-1,0); W^{2,q}(B(r)))$ be a solution to (E) for some $3 < q < +\infty$. Suppose there exists $r_0 \in (0,r)$ such that*

$$\limsup_{t\to 0}(-t)\|\nabla u(t)\|_{L^\infty(B(r_0))} < 1.$$

Then, $\limsup_{t\to 0}\|u(t)\|_{W^{2,q}(B(\rho))} < +\infty$ *for all* $\rho \in (0,r_0)$.

In a recent paper [14] the Type I condition of the above theorems is replaced by the condition involving the Hessian of the pressure as follows.

Theorem 4.8 *Let $(u,p) \in C^1(\mathbb{R}^3 \times (-1,0))$ be a solution of the Euler equation (E) with $u \in C([-1,0); W^{2,q}(\mathbb{R}^3))$, for some $q > 3$. If*

$$\limsup_{t\to 0}(-t)^2\|D^2 p(t)\|_{L^\infty} < 1,$$

then $\limsup_{t\to 0}\|u(t)\|_{W^{2,q}} < +\infty$.

This is also localized in the same paper [14].

Theorem 4.9 *Let $(u,p) \in C^1(B(x_0,\rho) \times (-\rho,0))$ be a solution to (E) with $u \in C([-\rho,0); W^{2,q}(B(x_0,\rho))) \cap L^\infty(0-\rho,0; L^2(B(x_0,\rho)))$ for some $q \in (3,\infty)$. If*

$$\int_{-\rho}^0 \|u(t)\|_{L^\infty(B(x_0,\rho))}dt < +\infty$$

and

$$\limsup_{t\to 0}(-t)^2\|D^2 p(t)\|_{L^\infty(B(x_0,\rho))} < 1,$$

then for all $r \in (0, \rho)$ we have

$$\limsup_{t \to 0} \|u(t)\|_{W^{2,q}(B(x_0,r))} < +\infty.$$

The following is a refined version of the above theorems [10, 15], considering also the sign condition for the Hessian of the pressure. We use the same notations as Proposition 3.11.

Theorem 4.10 *Let $(u, p) \in C^1(\mathbb{R}^3 \times (-1, 0))$ be a solution of the Euler equation (E) with $u \in C([-1, 0); W^{2,q}(\mathbb{R}^3))$, for some $q > 3$. Suppose the following holds. If either*

$$\limsup_{t \to 0} (-t)^2 \|[\zeta \cdot P\xi]_-(t)\|_{L^\infty} < 1,$$

or

$$\limsup_{t \to 0} (-t)^2 \|[|S\xi|^2 - 2\alpha^2 - \rho]_+(t)\|_{L^\infty} < 1,$$

then $\limsup_{t \to 0} \|u(t)\|_{W^{2,q}} < +\infty$.

The following is a localized version of the above theorem.

Theorem 4.11 *Let $(u, p) \in C^1(B(x_0, r) \times (-r, 0))$ be a solution to (E) with $u \in C([-r, 0); W^{2,q}(B(x_0, r))) \cap L^\infty(-r, 0; L^2(B(x_0, r)))$ for some $q \in (3, \infty)$. We suppose*

$$\int_{-r}^{0} \|u(t)\|_{L^\infty(B(x_0,r))} dt < +\infty.$$

If either

$$\limsup_{t \to 0} (-t)^2 \|[\zeta \cdot P\xi]_-(t)\|_{L^\infty(B(x_0,r))} < 1,$$

or

$$\limsup_{t \to 0} (-t)^2 \|[|S\xi|^2 - 2\alpha^2 - \rho]_+(t)\|_{L^\infty(B(x_0,r))} < 1,$$

then for all $\varepsilon \in (0, r)$ $\limsup_{t \to 0} \|u(t)\|_{W^{2,q}(B(x_0,\varepsilon))} < +\infty$.

5 Type I Blow-Up and the Energy Concentrations

Although we cannot exclude the possibility of Type I blow-up, we shall show in this section that under Type I condition the energy concentration in the form of atomic measure cannot happen at the blow-up time. Energy concentration in atomic form means that there exists an atomic measure μ (i.e. $\mu(\{x\}) > 0$ for some $x \in \mathbb{R}^3$) such that

$$|u(\cdot, t)|^2 dx \rightharpoonup \mu \quad \text{as} \quad t \to 0$$

in the sense of measure. A typical example is

$$|u(\cdot, t)|^2 dx \rightharpoonup \sum_{k=1}^{\infty} C_k \delta_{x_k}.$$

DSS singularity in the energy conserving scale is an example of Type I blow-up with one point energy conservation. Removing this scenario has been open. Concentration phenomena in the other equations are well studied. For example for the nonlinear Schödinger equations blow-up happens with L^2 norm concentration, while in the chemotaxis equations the blow-up occurs with L^1 norm concentration.

We first remove one point energy concentration under Type I. In the case $u \in L^\infty(-1, 0; L^2(\mathbb{R}^3))$, we can show that there exists a unique measure $\sigma \in \mathcal{M}(\mathbb{R}^3)$ such that

$$|u(t)|^2 dx \to \sigma \quad weakly\text{-}* \quad in \quad \mathcal{M}(\mathbb{R}^3) \quad as \quad t \to 0. \tag{77}$$

Here, we first consider the case σ is equal to the Dirac measure $E_0 \delta_0$ for some constant $0 \leqslant E_0 < +\infty$. Under the Type I condition we can exclude such one-point concentration of the energy, namely we have the following [22].

Theorem 5.1 *Let $u \in L^\infty(-1, 0; L^2(\mathbb{R}^3))$ be a solution to the Euler equations. In addition, we assume that u satisfies the Type I blow-up condition (69) and (82) with $\sigma = E_0 \delta_0$ for some $0 \leqslant E_0 < +\infty$. Then $u \equiv 0$.*

In the proof of the above theorem we use several decay properties of the solution to the Euler equations with respect to the space and time variables as we approach the blow-up time. The decay estimate is actually obtained under following more general condition than (69)

$$\exists \mu \in \left[\frac{3}{5}, 1\right): \quad \sup_{t \in (-1,0)} (-t)^{\frac{5}{3}\mu} \|\nabla u(t)\|_{L^\infty} < +\infty. \tag{78}$$

The following lemma is one of the two key decay estimates used to prove Theorem 5.1.

Lemma 5.2 *Let $u \in L^2(-\infty, 0; L^2(\mathbb{R}^3)) \cap L^\infty_{loc}([-1, 0), W^{1,\infty}(\mathbb{R}^3))$ be a solution to the Euler equations satisfying (69) and (82) with $\sigma = E\delta_0$. Then for every $0 < \beta < 5$ there exists a constant C such that for every $t \in [-1, 0)$ it holds*

$$\int_{\mathbb{R}^3} |v(t)|^2 |x|^\beta dx \leqslant C(-t)^{\beta(1-\alpha)}. \tag{79}$$

(Sketch of the Proof) We first claim the estimate

$$\|u(t)\|_{L^\infty} \leqslant C(-t)^{-\frac{3}{5}}. \tag{80}$$

Indeed, by the Gagliardo-Nirenberg interpolation we obtain

$$(-t)^{\frac{3}{5}}\|u(t)\|_{L^\infty} \leqslant C(-t)^{\frac{3}{5}}\|u(t)\|_{L^2}^{\frac{2}{5}}\|\nabla u(t)\|_{L^\infty}^{\frac{3}{5}}$$
$$\leqslant C E(-1)^{\frac{2}{5}}\{(-t)\|\nabla u(t)\|_{L^\infty}\}^{\frac{3}{5}} < +\infty.$$

We first prove the decay estimate for $\beta = 1$. We multiply (E) by $u|x|u\eta_R$ for a smooth cut-off η_R supported on the ball B_R, and integrate both sides over $\mathbb{R}^3 \times (t, 0)$. Integrating by parts, using the assumption of L^2-energy concentration at $x = 0$ as $t \to 0$, we have

$$\frac{1}{2}\int_{\mathbb{R}^3}|u(t)|^2|x|\eta_R dx$$
$$= \int_t^0\int_{\mathbb{R}^3}\eta_R|u|^2 u \cdot \nabla|x|dxds + \int_t^0\int_{\mathbb{R}^3}\eta_R pu \cdot \nabla|x|dxds$$
$$+ \frac{1}{2}\int_{\mathbb{R}^3}|u(0^+)|^2|x|\eta_R dx + \{\text{terms vanishing as } R \to +\infty\}$$
$$\leqslant \int_t^0\int_{\mathbb{R}^3}|u(s)|^2 dx\,\|v(s)\|_{L^\infty}ds + \int_t^0\int_{\mathbb{R}^3}|p(s)||v(s)|dxds + o(1)$$
$$\leqslant CE(-1)^2\int_t^0(-s)^{-\frac{3}{5}}ds + \int_t^0\|p(s)\|_{L^2}\|v(s)\|_{L^2}ds + o(1).$$

For the pressure term estimate we use the Calderon-Zygmund inequality $\|p\|_{L^q} \leqslant \|u\|_{L^{2q}}^2$, which follows from the well-known velocity-pressure relation $p = R_j R_k (u_j u_k)$, and estimate

$$\int_t^0\|p(s)\|_{L^2}\|u(s)\|_{L^2}ds \leqslant C\int_t^0\||u(s)|^2\|_{L^2}\|u(s)\|_{L^2}ds$$
$$\leqslant C\int_t^0\|u(s)\|_{L^\infty}\|u(s)\|_{L^2}^2 ds \leqslant C\{E(-1)\}^2\int_t^0(-s)^{-\frac{3}{5}}ds.$$

Passing $R \to \infty$, the lemma is proved for $\beta = 1$:

$$\int_{\mathbb{R}^3}|u(t)|^2|x|dx \leqslant C(-t)^{\frac{2}{5}}.$$

For $\beta > 1$ we multiply (E) by $u|x|^\beta\eta_R$, and integrate by parts as the above, and use the induction argument. For the pressure estimate we use the following weighted Calderon-Zygmund inequality (A_p weight) [53]:

$$\int_{\mathbb{R}^3}|p(s)|^2|x|^\gamma dx \leqslant C\int_{\mathbb{R}^3}|u(s)|^4|x|^\gamma dx \leqslant C(-s)^{-\frac{6}{5}}\int_{\mathbb{R}^3}|u(s)|^2|x|^\gamma dx,$$

which holds true for all $0 \leqslant \gamma < 3$.   ∎

The following is the second decay estimate for the Proof of Theorem 5.1.

Lemma 5.3 *Let* $u \in L^2(-1, 0; L^2_\sigma(\mathbb{R}^3)) \cap L^\infty_{loc}([-1, 0), W^{1,\infty}(\mathbb{R}^3))$ *be a solution to the Euler equations satisfying* (69) *for some* $\mu \in [\frac{3}{5}, 1)$ *and* (82) *with* $\sigma_0 = E\delta_0$. *Then for all* $k \in \mathbb{N} \cup \{0\}$ *and* $0 < r < +\infty$ *it holds*

$$\|\mathbb{P}_r(v(t))\|^2_{L^2(B(r)^c)} \leqslant C_0^k 4^{k^2} (-t)^{(1-\mu)k} r^{-k} \quad \forall t \in (-1, 0),$$

where $\mathbb{P}_r$ *is the Helmholtz projection operator on* $B(r)$.

(Sketch of the proof of Theorem 5.1) We choose θ small enough: $0 < \theta < \frac{1}{5}$. For a solution u to the Euler equations we transform: $u \mapsto w$,

$$w(x, t) = u((-t)^\theta x, t).$$

Then, w solves the transformed Euler system,

$$w_t + \theta(-t)^{-1} x \cdot \nabla w + (-t)^{-\theta}(w \cdot \nabla) w = -\nabla \pi,$$
$$\nabla \cdot w = 0.$$

Using the two decay lemmas above, one can show that there exists $t_0 \in (-1, 0)$ such that
$$\nabla \times w(t) = 0 \quad \text{on} \quad B(1)^c \quad \forall t_0 < t < 0.$$

Transforming back to the original vorticity, $\omega(t) = \nabla \times u(t)$, we find

$$\text{supp}\, \omega(t) \subset B((-t)^\theta) \quad \forall t_0 < t < 0.$$

Since the measure of $\text{supp}\, \omega(t)$ is preserved due to the Cauchy formula,

$$\omega(X(a, t), t) = \nabla_a X(a, t)\omega_0(a),$$

we have
$$\text{meas}\{\text{supp}\, \omega(t_0)\} = \text{meas}\{\text{supp}\, \omega(t)\} \leqslant C(-t)^{3\theta} \to 0$$

as $t \to 0$. Whence, $\omega(t_0) \equiv 0$, and $u(t_0)$ is harmonic. Since $u(t_0) \in L^2(\mathbb{R}^3)$, we conclude that $u(t_0) \equiv 0$, and hence $u \equiv 0$, which is a contradiction. Therefore, one point energy concentration under the Type I condition is impossible.   ∎

As an immediate corollary of the above theorem we establish the following.

Corollary 5.4 *Let* $u \in L^\infty(-1, 0; L^2(\mathbb{R}^3)) \cap L^\infty_{loc}([-1, 0), W^{1,\infty}(\mathbb{R}^3))$ *be a DSS solution to the Euler equation, i.e. there exists* $\lambda > 1$ *such that*

$$u(x, t) = \lambda^{\frac{3}{2}} u(\lambda x, \lambda^{\frac{5}{2}} t) \quad \forall (x, t) \in \mathbb{R}^3 \times (-1, 0).$$

Then $u \equiv 0$.

For the proof we refer to [22].

Next, we shall use the blow-up argument to remove more general form of atomic concentration under local Type I condition. More specifically, we have the following.

Theorem 5.5 *Let $u \in L^\infty(-1, 0; L^2(\mathbb{R}^3)) \cap L^\infty_{loc}([-1, 0); W^{1, \infty}(\mathbb{R}^3))$ be a solution of the Euler equations satisfying the Type I condition,*

$$\sup_{t \in (-1, 0)} (-t) \|\nabla u(t)\|_{L^\infty} < +\infty.$$

Suppose there exists $\sigma_0 \in \mathcal{M}(\mathbb{R}^3)$ such that

$$|u(t)|^2 dx \to \sigma_0 \quad as \ \ t \to 0^-.$$

Then, σ_0 is non-atomic.

We first recall the notion of suitable weak solution (u, p) of (E), a weak solution satisfying the local energy inequality:

$$\int_{\mathbb{R}^3} |u(t)|^2 \phi dx \leqslant \int_{\mathbb{R}^3} |u(s)|^2 \phi dx + \int_s^t \int_{\mathbb{R}^3} (|v|^2 + 2p) u \cdot \nabla \phi dx d\tau$$

for all $\phi \in C_c^\infty(\mathbb{R}^3)$ and for a.e. $-1 \leqslant t < s < 0$. Below we denote the 'parabolic cylinder' consistent with the energy conserving scale by $Q(R) := B(R) \times (-R^{5/2}, 0)$. Then we establish the following criterion of energy non-concentration in terms of a Morrey norm.

Theorem 5.6 *We set the cylinder $Q(R) = B(R) \times (-R^{\frac{5}{2}}, 0)$. Let $u \in L^\infty(-R^{\frac{5}{2}}, 0;$ $L^2(B(R))) \cap L^3(Q(R))$ be a local suitable weak solution to (E) such that the local energy inequality is satisfied. Furthermore, we assume that*

$$\sup_{0 < r \leqslant R} r^{-1} \|u\|^3_{L^3(Q(r))} < +\infty, \quad \liminf_{r \to 0^+} r^{-1} \|u\|^3_{L^3(Q(r))} = 0. \tag{81}$$

Then, there is no energy concentration at the point $x = 0$ as $t \to 0$.

Remark In [52] Shvydkoy showed that if $u \in L^q(-1, 0; L^\infty(\Omega)) \cap L^\infty(-1, 0; L^2(\Omega))$, $q = \frac{5}{3}$, is a suitable weak solution, then there is no atomic concentration in Ω. This actually follows from the above theorem immediately. Let $Q(r) \subset \Omega \times (-1, 0)$. Then

$$r^{-1}\|u\|^3_{L^3(Q(r))} = r^{-1}\int_{-r^{\frac{5}{2}}}^{0}\int_{B(r)}|u|^3 dx dt$$

$$\leqslant \|u\|^2_{L^\infty(-1,0;L^2(\Omega))} r^{-1}\int_{-r^{\frac{5}{2}}}^{0}\|u\|_{L^\infty(B(r))} dt$$

$$\leqslant \|u\|^2_{L^\infty(-1,0;L^2(\Omega))}\left(\int_{-r^{\frac{5}{2}}}^{0}\|u\|^{\frac{5}{3}}_{L^\infty(B(r))} dt\right)^{\frac{3}{5}} \to 0$$

as $r \to 0$.

(Sketch of the proof of Theorem 5.6) We shall use the blow-up argument for the proof of the theorem. Let us first note the following interpolation inequality,

$$r^{-1}\|u\|^3_{L^3(Q(r))} \leqslant C K_0 r^{-\frac{5}{2}}\|u\|^2_{L^2(Q(r))} + C K_0^{\frac{1}{2}} K_1^{\frac{3}{2}}\left(r^{-\frac{5}{2}}\|u\|^2_{L^2(Q(r))}\right)^{\frac{1}{2}}, \qquad (82)$$

where we set

$$K_0 := \|u(t)\|_{L^\infty(-R^{5/2},0);L^2(B(R))}, \quad K_1 := \sup_{t\in(-R^{\frac{5}{2}},0)} (-t)\|\nabla u(t)\|_{L^\infty(B(R))},$$

which are bounded constants by the hypothesis. Suppose there exists an atomic concentration. Then Theorem 5.6, combined with the above interpolation inequality (82) implies that there exists $\varepsilon > 0$ and a sequence $r_k \to 0$ such that

$$r_k^{-\frac{5}{2}}\|u\|^2_{L^2(Q(r_k))} \geqslant \varepsilon \quad \forall k \in \mathbb{N}.$$

We define a (blow-up) sequence

$$u_k(x,t) = r_k^{\frac{3}{2}} u(r_k x, r_k^{\frac{5}{2}} t), \quad k \in \mathbb{N}.$$

Using Type I condition and the energy conservation, we can deduce the following uniform bound for $\{u_k\}$,

$$\|u_k\|_{L^\infty(-1,0;L^2(\mathbb{R}^3))} + \|u_k\|_{L^3((-1,0);\dot{W}^{\theta,3}(\mathbb{R}^3))} \leqslant C$$

for all $0 < \theta < \frac{1}{3}$. Here, we use the following norm for the fractional derivatives (Sobolev-Slobodeckij semi-norm) in $\mathbb{R}^3$,

$$|f|_{\dot{W}^{\theta,p}} := \left(\int_{\mathbb{R}^3}\int_{\mathbb{R}^3}\frac{|f(x)-f(y)|^p}{|x-y|^{\theta p+3}} dx dy\right)^{\frac{1}{p}}.$$

Taking the limit for a sub-sequence (by compactness lemma), one can construct a non-trivial suitable weak solution to (E),

$$u^* \in L^\infty(-1, 0; L^2_\sigma(\mathbb{R}^3)) \cap L^3([-1, 0); \dot{W}^{\theta, 3}(\mathbb{R}^3))$$

satisfying the following 'weaker-norm version' of local Type I condition

$$\sup_{r \in (0, R)} \frac{1}{r^{1-3\theta}} \int_{-r^{\frac{5}{2}}}^{0} |u^*(t)|^3_{\dot{W}^{\theta, 3}(B(r))} dt < +\infty.$$

Indeed, we have the following another interpolation inequality:

$$\sup_{r \in (0, R)} \frac{1}{r^{1-3\theta}} \int_{-r^{\frac{5}{2}}}^{0} |u(t)|^3_{\dot{W}^{\theta, 3}(B(r))} dt \leqslant C \sup_{r \in (0, R)} r^{-1} \|u\|^3_{L^3(Q(r))}$$

$$+ C \sup_{-R^{\frac{5}{2}} < t < 0} (-t)^3 \|\nabla u(t)\|^3_{L^\infty(B(R))} < +\infty$$

by (82) and the Type I condition respectively. Moreover, for such limiting solution u^* one can choose a sequence of time $\{t_k\} \subset [-1, 0)$ and a positive constant $c_0 > 0$ such that

$$|u^*(x, t_k)|^2 dx \rightharpoonup C_0 \delta_0 \quad \text{as} \quad k \to +\infty$$

in the sense of measure, namely one point concentration in $\mathbb{R}^3$ for blow-up limiting solution. Our previous exclusion theorem for one point energy concentration in $\mathbb{R}^3$ with Type I blow-up condition implies $C_0 = 0$, namely no atomic concentration. ∎

6 The Boussinesq Equations

We consider the Boussinesq equations in the space time cylinder $\mathbb{R}^2 \times (0, \infty)$

$$(B) \begin{cases} \partial_t u + (u \cdot \nabla)u = e_2\theta - \nabla p, \\ \partial_t \theta + (u \cdot \nabla)\theta = 0, \\ \nabla \cdot u = 0, \quad u(x, 0) = u_0(x), \quad \theta(x, 0) = \theta_0(x) \end{cases}$$

where $u = (u_1(x, t), u_2(x, t))$, $(x, t) \in \mathbb{R}^2 \times (0, +\infty)$ is the fluid velocity, while $\theta = \theta(x, t)$ represents the temperature of the fluid, and $e_2 = (0, 1)$. The system (B) is a fundamental system of equations describing the motion of atmosphere (see e.g. [49, 50]). Besides its importance in application to the atmospheric sciences another reason why the Boussinesq equation attracted many mathematicians is that the system (B) has strong similarity to the 3D axisymmetric Euler equations, thus providing a good model problem for the Euler equations. To see this resemblance between the two equations more closely we consider the following vorticity equation, obtained by operating $\nabla^\perp \cdot$ on the first equation of (B):

$$\partial_t \omega + u \cdot \nabla \omega = \partial_1 \theta, \quad \omega = \partial_1 u_2 - \partial_2 u_1. \tag{83}$$

Setting $\Theta = (ru^\theta)^2$ and $W = \frac{\omega^\theta}{r}$, the axisymmetric Euler system (64), (62) and (61) can be written as

$$\begin{cases} W_t + u^r \partial_r W + u^3 \partial_3 W = \dfrac{1}{r^4} \partial_3 \Theta, \\ \Theta_t + u^r \partial_r \Theta + u^3 \partial_3 \Theta = 0, \quad \partial_r(ru^r) + \partial_3(ru^3) = 0. \end{cases} \tag{84}$$

Therefore, if we consider the system (84) off the axis region($r > 0$) the system (B) has the almost same structure as (84) with the correspondence

$$(x_1, x_2) \Leftrightarrow (r, x_3), \quad (u_1, u_2) \Leftrightarrow (u^r, u^\theta), \quad (\omega, \theta) \Leftrightarrow (W, \Theta).$$

Let us consider the particle trajectory $X(\alpha, t)$ generated by $u = u(x, t)$. Then, the second equation of (B) implies the conservation

$$f(\theta(X(\alpha, t), t)) = f(\theta_0(\alpha)) \quad \forall f \in C^1(\mathbb{R}).$$

The following proposition shows that a certain quantity, which corresponds to the Helicity of the 3D Euler equations, has localized conservation law.

Proposition 6.1 *Let f be a $C^1(\mathbb{R})$, and (u, θ) be a smooth solution to (B), and C_t, $t \geqslant 0$ be a level curve of $\theta(\cdot, t)$. Then,*

$$\oint_{C_t} u \cdot \nabla^\perp f(\theta) ds = \oint_{C_0} u_0 \cdot \nabla^\perp f(\theta_0) ds \quad \forall t > 0. \tag{85}$$

(Proof) From the second equation of (B) we have

$$\frac{D}{Dt} \nabla^\perp \theta = (\partial_t + u \cdot \nabla) \nabla^\perp \theta = \nabla^\perp \theta \cdot \nabla u. \tag{86}$$

Let

$$C_0 = \{\gamma(s) \,:\, \theta_0(\gamma(s)) = \lambda, s \in [0, 1], \gamma(0) = \gamma(1)\}$$

be a closed level curve for θ_0. Define

$$C_t = X(C_0, t) = \{X(\gamma(s), t) \,:\, 0 \leqslant s \leqslant 1\}.$$

Then, for any $f \in C^1(\mathbb{R})$, we have

$$\frac{d}{dt} \int_0^1 u(X(\gamma(s), t), t) \cdot \nabla^\perp f(\theta(X(\gamma(s), t), t)) ds$$

$$= f'(\lambda) \int_0^1 \frac{D}{Dt} u(X(\gamma(s), t), t) \cdot \nabla^\perp \theta(X(\gamma(s), t), t) ds$$

$$+ f'(\lambda) \int_0^1 u(X(\gamma(s), t), t) \cdot \frac{D}{Dt} \nabla^\perp \theta(X(\gamma(s), t), t) ds$$

$$= -f'(\lambda) \int_0^1 \nabla p(X(\gamma(s), t), t) \cdot \nabla^\perp \theta(X(\gamma(s), t), t) ds$$

$$+ f'(\lambda) \int_0^1 \theta(X(\gamma(s), t), t) e_2 \cdot \nabla^\perp \theta(X(\gamma(s), t), t) ds$$

$$+ f'(\lambda) \int_0^1 u(X(\gamma(s), t), t) \cdot \nabla^\perp \theta(X(\gamma(s), t), t) \cdot \nabla u(X(\gamma(s), t), t) ds$$

$$= K_1 + K_2 + K_3.$$

We compute each term separately. First,

$$K_1 = -f'(\lambda) \int_0^1 \nabla p(X(\gamma(s), t), t) \cdot \frac{\partial}{\partial s} X(\gamma(s), t) ds$$

$$= -f'(\lambda) \int_0^1 \frac{\partial}{\partial s} p(X(\gamma(s), t), t) ds = 0.$$

Second,

$$K_2 = f'(\lambda) \lambda e_2 \cdot \int_0^1 \frac{\partial X(\gamma(s), t), t)}{\partial s} ds = 0.$$

Finally,

$$K_3 = f'(\lambda) \int_0^1 u(X(\gamma(s), t), t) \cdot \left(\frac{\partial}{\partial s} X(\gamma(s), t) \cdot \nabla \right) u(X(\gamma(s), t), t) ds$$

$$= \frac{1}{2} f'(\lambda) \int_0^1 \left(\frac{\partial}{\partial s} X(\gamma(s), t) \cdot \nabla \right) |u(X(\gamma(s), t), t)|^2 ds$$

$$= \frac{1}{2} f'(\lambda) \int_0^1 \frac{\partial}{\partial s} |u(X(\gamma(s), t), t)|^2 ds = 0.$$

Combining the calculations for K_1, K_2, K_3 above, we find that

$$\frac{d}{dt} \int_0^1 u(X(\gamma(s), t), t) \cdot \nabla^\perp f(\theta(X(\gamma(s), t), t)) ds = 0.$$

This completes the proof of the proposition. ∎

Regarding the Cauchy problem for the system (B) for the initial data in $H^k(\mathbb{R}^2)$, $k > 2$, the local-in-time existence of solution and the Beale-Kato-Majda type blow-up criterion are first obtained in [16].

Theorem 6.2 *Let* $(u_0, \theta_0) \in H^k(\mathbb{R}^2)$, $k > 2$. *Then, there exists* $T = T(\|u_0\|_{H^k}, \|\theta_0\|_{H^k})$ *such that a unique solution* $u \in C([0, T); H^k(\mathbb{R}^2))$ *exists. Furthermore,*

$$\limsup_{t \to T}(\|u(t)\|_{H^k} + \|\theta(t)\|_{H^k}) = +\infty \quad \text{if and only if}$$

$$\int_0^T \|\nabla\theta(t)\|_{L^\infty}dt = +\infty. \tag{87}$$

The finite time blow-up question for the Boussinesq system with a smooth initial data is also a wide-open problem. We mention that for domain with cusp singularity finite time blow-up at the boundary point is obtained recently in [37], and also in [27] the authors show singularity on the boundary point of a cylinder. Our main concern here is the possibility of interior singularity in the whole domain of $\mathbb{R}^2$. It is also worth mentioning that if we add viscosity term to either one of the velocity or the temperature equations of (B), the finite time blow-up question was posed by Moffatt in [51] as one of the millennium problems in the fluid mechanics, for which there was a partial result due to Córdoba, Fefferman and LLave in [34], removing "squirt" singularities. The problem is fully resolved in [11], which shows that there exists no finite time singularities in this partially viscous case.

A similar result to Theorem 6.2 in the setting of the Hölder space is proved in [17]. For the BKM type criterion an improvement of (87) has been obtained in [25] as follows.

Theorem 6.3 *Let* $(u, \theta) \in C([0, T); W^{2,q}(\mathbb{R}^2))$, $q > 2$, *be a solution of (B). If*

$$\int_0^T (T - t)\|\nabla\theta(t)\|_{L^\infty}dt < +\infty, \tag{88}$$

then there exists no blow-up at $t = T$, *and thus both* u *and* θ *belong to* $C([0, T]; W^{2,q}(\mathbb{R}^2))$.

(Proof) For convenience we shift in time so that $[0, T] \mapsto [-1, 0]$.
Step (i) We first show that

$$\int_{-1}^0 \|\omega(t)\|_{L^\infty}dt + \int_{-1}^0 (-t)\|\nabla\theta(t)\|_{L^\infty}dt < +\infty \tag{89}$$

implies no blow-up at $t = 0$. Let $q > 2$. We apply the operator ∂_i to the vorticity equation, multiplying the resultant equation by $\partial_i\omega|\nabla\omega|^{q-1}$, and integrating it over

$\mathbb{R}^2$. Then, after the integration by parts and using the Hölder inequality, we are led to

$$\frac{d}{dt}\|\nabla\omega\|_{L^q} \leqslant \|\nabla u\|_{L^\infty}\|\nabla\omega\|_{L^q} + \|\nabla^2\theta\|_{L^q}$$
$$= \|\nabla u\|_{L^\infty}\|\nabla\omega\|_{L^q} + (-t)^{-1}(-t)\|\nabla^2\theta\|_{L^q}. \tag{90}$$

Next, we apply the operator $\partial_i\partial_j$ to both sides of the θ equation, multiply both sides by $\partial_i\partial_j\theta|\nabla^2\theta|^{q-2}$, and sum over $i, j = 1, 2, 3$, and the integrate it over $\mathbb{R}^2$. This, applying the integration by part and the Hölder inequality, yields the following inequality

$$\frac{d}{dt}\|\nabla^2\theta\|_{L^q} \leqslant 2\|\nabla u\|_{L^\infty}\|\nabla^2\theta\|_{L^q} + \|\nabla\theta\|_{L^\infty}\|\nabla^2 u\|_{L^q}. \tag{91}$$

Multiplying both sides of (91) by $(-t)$, we see that

$$\frac{d}{dt}(-t)\|\nabla^2\theta\|_{L^q} + \|\nabla^2\theta\|_{L^q}$$
$$\leqslant 2\|\nabla u\|_{L^\infty}(-t)\|\nabla^2\theta\|_{L^q} + (-t)\|\nabla\theta\|_{L^\infty}\|\nabla^2 u\|_{L^q}$$
$$\leqslant 2\|\nabla u\|_{L^\infty}(-t)\|\nabla^2\theta\|_{L^q} + c_{cz}(-t)\|\nabla\theta\|_{L^\infty}\|\nabla\omega\|_{L^q}. \tag{92}$$

Now define

$$\Psi(t) := \|\nabla\omega\|_{L^q} + (-t)\|\nabla^2\theta\|_{L^q}, \quad t \in (-1, 0).$$

Adding the last two inequalities (90) and (92), we are led to

$$\Psi' \leqslant \left(2\|\nabla u(t)\|_{L^\infty} + (-t)^{-1} + c_{cz}(-t)\|\nabla\theta(t)\|_{L^\infty}\right)\Psi. \tag{93}$$

By means of the logarithmic Sobolev embedding of the Beale-Kato-Majda type, we find

$$\|\nabla v(t)\|_{L^\infty} \leqslant C\left\{1 + \|\omega(t)\|_{L^\infty}\log(e + \|\nabla^2 u(t)\|_{L^q})\right\}$$
$$\leqslant C\left\{1 + \|\omega(t)\|_{L^\infty}\log(e + \Psi(t))\right\}. \tag{94}$$

Inserting (94) into (93), it follows

$$\Psi' \leqslant \left\{C\left[1 + (\|\omega(t)\|_{L^\infty} + (-t)\|\nabla\theta(t)\|_{L^\infty})\log(e + \Psi(t))\right] + (-t)^{-1}\right\}\Psi(t). \tag{95}$$

Setting $y(t) = \log(e + \Psi(t))$, we infer from (95) the differential inequality

$$y' \leqslant Ca(t)y + C(-t)^{-1}, \qquad a(t) = \|\omega(t)\|_{L^\infty} + (-t)\|\nabla\theta(t)\|_{L^\infty} \tag{96}$$

which can be solved as

$$y(t) = \log(e + \Psi(t))$$

$$\leqslant y(t_0)e^{C\int_{t_0}^t a(s)ds} + C\int_{t_0}^t (-s)^{-1} e^{c\int_s^t a(\tau)d\tau} ds. \qquad (97)$$

We now choose t_0 so that $e^{C\int_{t_0}^0 a(s)ds} < 2$. Then, (97) implies

$$\log(e + \Psi(t)) \leqslant C\log(e + \Psi(t_0)) + C\log(-1/t) \qquad \forall t \in (t_0, 0), \qquad (98)$$

where $c > 2$ is another constant. From θ-equation of (B) we have immediately

$$\frac{\partial}{\partial t}|\nabla\theta| + (u \cdot \nabla)|\nabla\theta| \leqslant |\nabla u||\nabla\theta|. \qquad (99)$$

Let $t \in (-1, 0)$ be arbitrarily chosen but fixed. Let $x_0 \in \mathbb{R}^2$. By $X(x_0, t)$ we denote the trajectory of the particle which is located at x_0 at time $t = t_0$, defined by the following ODE

$$\frac{dX(x_0, t)}{dt} = u(X(x_0, t), t) \quad \text{in} \quad [-1, 0), \quad X(x_0, t_0) = x_0. \qquad (100)$$

Then, (99) can be written as

$$\frac{\partial}{\partial t}|\nabla\theta(X(x_0, t), t)| \leqslant |\nabla u(X(x_0, t), t)||\nabla\theta(X(x_0, t), t)|, \qquad (101)$$

which can be integrated along the trajectories as

$$|\nabla\theta(X(x_0, t), t)| \leqslant |\nabla\theta(x_0)| \exp\left(\int_{t_0}^t |\nabla u(X(x_0, s), s)|ds\right).$$

Therefore, we estimate, using (94) as

$$\|\nabla\theta(t)\|_{L^\infty} \leqslant \|\nabla\theta(t_0)\|_{L^\infty} \exp\left(\int_{t_0}^t \|\nabla u\|_{L^\infty}ds\right)$$

$$\leqslant \|\nabla\theta(t_0)\|_{L^\infty} \exp\left(C\int_{t_0}^t \{\|\omega(s)\|_{L^\infty} [\log(e + \Psi(t_0)) + \log(-1/s)] + 1\}ds\right)$$

$$\leqslant \|\nabla\theta(t_0)\|_{L^\infty} \times$$

$$\times \exp\left(C\{\log(e + \Psi(t_0)) + \log(-1/t)\}\int_{t_0}^t \|\omega(s)\|_{L^\infty}ds + c(t - t_0)\right).$$

$$(102)$$

Choosing $t_0 \in (-1, 0)$ so that $C \int_{t_0}^{0} \|\omega(s)\|_{L^\infty} ds < \frac{1}{2}$, we deduce from (102) that

$$\|\nabla\theta(t)\|_{L^\infty} \leqslant \|\nabla\theta(t_0)\|_{L^\infty} (e + \Psi(t_0))^C e^C (-t)^{-\frac{1}{2}} \quad \forall t \in (t_0, 0).$$

Therefore, $\int_{-1}^{0} \|\nabla\theta\|_{L^\infty} dt < +\infty$. Applying the well-known blow-up criterion in [5], we obtain the desired result of (89).

Step (ii) Here we show the estimate:

$$\int_{-1}^{t} \|\omega(s)\|_{L^\infty} ds + \int_{-1}^{t} (-s)\|\nabla\theta(s)\|_{L^\infty} ds$$

$$\leqslant \|\omega(-1)\|_{L^\infty} + 2\int_{-1}^{0} (-s)\|\nabla\theta(s)\|_{L^\infty} ds < +\infty, \tag{103}$$

thus finishing the proof, combining this with (89). We recall the vorticity equation from (B).

$$\partial_t \omega + u \cdot \nabla\omega = \partial_1\theta \quad \text{in} \quad \mathbb{R}^2 \times [-1, 0), \tag{104}$$

where $\omega = \partial_1 u_2 - \partial_2 u_1$. Using the particle trajectories with $X(x_0, -1) = x_0$ as the above, we have from (104)

$$\frac{d}{dt}|\omega(X(x_0, t), t)| \leqslant |\partial_1\theta(X(x_0, t), t)| \quad \text{in} \quad [-1, 0), \tag{105}$$

which implies that

$$\|\omega(s)\|_{L^\infty} \leqslant \|\omega(-1)\|_{L^\infty} + \int_{-1}^{s} \|\partial_1\theta(\tau)\|_{L^\infty} d\tau. \tag{106}$$

Integrating both sides of (106) over $[-1, t), t \in (-1, 0)$ with respect to s, and applying integration by parts, we get

$$\int_{-1}^{t} \|\omega(s)\|_{L^\infty} ds \leqslant (1 + t)\|\omega(-1)\|_{L^\infty} + \int_{-1}^{t} \int_{-1}^{s} \|\partial_1\theta(\tau)\|_{L^\infty} d\tau ds$$

$$= (1 + t)\|\omega(-1)\|_{L^\infty} + \int_{-1}^{t} \left\{ \frac{d}{ds}(s) \int_{-1}^{s} \|\partial_1\theta(\tau)\|_{L^\infty} d\tau \right\} ds$$

$$= (1 + t)\|\omega(-1)\|_{L^\infty} + \int_{-1}^{t} (-s)\|\partial_1\theta(s)\|_{L^\infty} ds + t \int_{-1}^{t} \|\partial_1\theta(s)\|_{L^\infty} ds$$

$$\leqslant \|\omega(-1)\|_{L^\infty} + \int_{-1}^{t} (-s)\|\partial_1\theta(s)\|_{L^\infty} ds.$$

∎

The above theorem has been also localized in [23] as follows.

Theorem 6.4 *Let $B(r) \subset \mathbb{R}^2$ be the unit ball, $2 < q < +\infty$, and*

$$(u, \theta) \in C([0, T); W^{2,q}(B(r))) \times C([0, T); W^{2,q}(B(r)))$$

be a solution to (B). Suppose that

$$u \in L^\infty(0, T; L^2(B(r)))$$

and

$$\int_0^T (T - t)\|\nabla\theta(t)\|_{L^\infty(B(r))}dt < +\infty, \quad \int_0^T \|u(t)\|_{L^\infty(B(r))}dt < +\infty.$$

Then $u, \theta \in C([0, T], W^{2,q}(B(\rho)))$ for all $0 < \rho < r$.

The blow-up criterion in terms of the Hessian of the pressure is also recently obtained as follows. For a solution (u, p, θ) of the system (B) let us introduce the $\mathbb{R}^{2\times2}$-valued functions $U = (\partial_i u_j)$ and $P = (\partial_i \partial_j p)$. For the vector field $\nabla^\perp \theta = (-\partial_2\theta, \partial_1\theta)$ we define the direction vectors

$$\xi = \nabla^\perp\theta/|\nabla^\perp\theta|, \quad \zeta = U\nabla^\perp\theta/|U\nabla^\perp\theta|.$$

We note that contrary to the case of Euler equations U is not the symmetric part of the velocity gradient matrix. Then, the following blow-up criterion in terms of the Hessian of the pressure is proved in [14].

Theorem 6.5 *Let $(u, p) \in C^1(\mathbb{R}^2 \times (0, T))$ be a solution of the Boussinesq equation (B) with $u \in C([0, T); W^{2,q}(\mathbb{R}^2))$, for some $q > 2$. Suppose the following holds. Either*

$$\int_0^T (T - t) \exp\left(\int_0^t \int_0^s \|[\zeta \cdot P\xi]_-(\tau)\|_{L^\infty}d\tau ds\right) dt < +\infty,$$

or

$$\limsup_{t \to T} (T - t)^2 \|[\zeta \cdot P\xi]_-(t)\|_{L^\infty} < 2.$$

Then $\limsup_{t \to T} \|u(t)\|_{W^{2,q}} < +\infty$.

Note the relaxed smallness condition for the nonexistence of Type I blow-up compared to the case of 3D Euler equations. This is due to the extra factor, $(T - t)$ in the integral $\int_0^T (T - t)\|\nabla^\perp\theta(t)\|_{L^\infty}dt < +\infty$ in Theorem 6.3.

(Proof of the first part of Theorem 6.5) Let (u, p, θ) be a solution of (B), which belongs to $C^1(\mathbb{R}^2 \times (0, T))$. We first claim the following formula.

$$D_t|U\nabla^\perp\theta| = -\zeta \cdot P\nabla^\perp\theta. \tag{107}$$

Indeed, ∇ on the first equation of (B), we find

$$D_t U + U^2 = -P + \nabla(\theta e_2).$$

Taking $\nabla^\perp$ on the second equation of (B), we obtain

$$D_t \nabla^\perp \theta = U \nabla^\perp \theta.$$

Let us compute

$$
\begin{aligned}
D_t^2 \nabla^\perp \theta &= D_t U \nabla^\perp \theta + U D_t \nabla^\perp \theta \\
&= -U^2 \nabla^\perp \theta - P \nabla^\perp \theta + U^2 \nabla^\perp \theta + \nabla^\perp \theta \cdot \nabla(\theta e_2) \\
&= -P \nabla^\perp \theta,
\end{aligned}
\tag{108}
$$

where we use the fact

$$\nabla^\perp \theta \cdot \nabla(\theta e_2) = 0.$$

We multiply (110) by $D_t \nabla^\perp \theta$ to have

$$
\begin{aligned}
|D_t \nabla^\perp \theta| D_t |D_t \nabla^\perp \theta| = \frac{1}{2} D_t \left(|D_t \nabla^\perp \theta|^2 \right) &= D_t \nabla^\perp \theta \cdot D_t^2 \nabla^\perp \theta \\
&= -U \nabla^\perp \theta \cdot P \nabla^\perp \theta.
\end{aligned}
\tag{109}
$$

the left-hand side of which can be re written as

$$\frac{1}{2} D_t |U \nabla^\perp \theta|^2 = |U \nabla^\perp \theta| D_t |U \nabla^\perp \theta|.$$

Hence, dividing the both sides of (109) by $|U \nabla^\perp \theta|$, we obtain the formula (107), and the claim is proved.

Now, integrating (107) along the trajectory for $t \in [0, s]$, we obtain

$$
\begin{aligned}
\frac{\partial}{\partial s} |\nabla^\perp \theta(X(\alpha, s), s)| &\leqslant \left| \frac{\partial}{\partial s} \nabla^\perp \theta(X(\alpha, s), s) \right| \\
&= |(D_s \nabla^\perp \theta)(X(\alpha, s), s)| = |U \nabla^\perp \theta(X(\alpha, s), s)| \\
&= |S_0(\alpha) \omega_0(\alpha)| - \int_0^s (\zeta \cdot P\xi)(X(\alpha, \tau), \tau) |\omega(X(\alpha, \tau), \tau| d\tau.
\end{aligned}
$$

After integrating this again with respect to s over $[0, t]$, we find

$$
\begin{aligned}
|\nabla^\perp \theta(X(\alpha, t), t)| \leqslant\ & |\nabla^\perp \theta_0(\alpha)| + |\nabla^\perp \theta_0(\alpha) \cdot \nabla u_0(\alpha)| t \\
& + \int_0^t \int_0^s [\zeta \cdot P\xi]_-(X(\alpha, \tau), \tau) |\nabla^\perp \theta(X(\alpha, \tau), \tau)| d\tau ds.
\end{aligned}
$$

Thanks to Theorem 3.8 we find

$$|\nabla^\perp\theta(X(\alpha,t),t)| \leqslant (|\nabla^\perp\theta_0(\alpha)| + |\nabla^\perp\theta_0 \cdot \nabla u_0(\alpha)|t) \times$$
$$\times \exp\left(\int_0^t \int_0^s [\zeta \cdot P\xi]_-(X(\alpha,\tau),\tau)d\tau ds\right).$$

Taking the supremum over $a \in \mathbb{R}^2$, and integrating it with respect to t over $[0, T]$ after multiplying by $T - t$, we obtain

$$\int_0^T (T-t)\|\nabla^\perp\theta(t)\|_{L^\infty}dt \leqslant (\|\nabla^\perp\theta_0\|_{L^\infty} + \|\nabla^\perp\theta_0 \cdot \nabla u_0\|_{L^\infty}T) \times$$
$$\times \int_0^T (T-t)\exp\left(\int_0^t \int_0^s \|[\zeta \cdot P\xi(\tau)]_-\|_{L^\infty}d\tau ds\right)dt.$$

Applying the blow-up criterion of Theorem 6.3, we obtain the desired conclusion. ∎

The following is a localized version of Theorem 6.5.

Theorem 6.6 Let $(u, p) \in C^1(B(x_0, r) \times (T - r, T))$ be a solution to (E) with $u \in C([T - r, T); W^{2,q}(B(x_0, r))) \cap L^\infty(T - r, T; L^2(B(x_0, r)))$ for some $q \in (2, \infty)$. Let us assume

$$\int_{T-r}^T \|u(t)\|_{L^\infty(B(x_0,r))}dt < +\infty. \tag{110}$$

If either

$$\int_{T-r}^T (T-t)\exp\left(\int_0^t \int_0^s \|[\zeta \cdot P\xi]_-(\tau)\|_{L^\infty(B(x_0,r))}d\tau ds\right)dt < +\infty,$$

or

$$\limsup_{t \to T}(T-t)^2\|[\zeta \cdot P\xi]_-(t)\|_{L^\infty(B(x_0,r))} < 2,$$

then for all $\varepsilon \in (0, r)$ $\limsup_{t \to T}\|u(t)\|_{W^{2,q}(B(x_0,\varepsilon))} < +\infty$.

For the proof we first show that the condition (110) implies that the mapping $t \mapsto X(\alpha, t)$ belongs to $C([T - r, T]; \mathbb{R}^3)$ for all $\alpha \in B(x_0, r)$. Then, the other part of the proof follows by applying the continuity argument. For more details we refer to [15].

Acknowledgements This research was supported partially by NRF grant 2021R1A2C1003234.

References

1. C. Bardos, E.S. Titi, Euler equations for an ideal incompressible fluid. Russ. Math. Surv. **62**(3), 409–451 (2007)
2. J.T. Beale, T. Kato, A. Majda, Remarks on the breakdown of smooth solutions for the 3-D Euler equations. Comm. Math. Phys. **94**, 61–66 (1984)
3. D. Chae, *Incompressible Euler Equations: Mathematical Theory, Encyclopedia of Mathematical Physics*, vol. 3 (Elsevier/Academic Press, Oxford, 2006), pp. 10–17
4. D. Chae, *Incompressible Euler Equations: The Blow-Up Problem and Related Results, Handbook of Differential Equations: Evolutionary Equations*, vol. IV (Elsevier/North-Holland, Amsterdam, 2008), pp. 1–55
5. D. Chae, Nonexistence of self-similar singularities for the 3D incompressible Euler equations. Comm. Math. Phys. **273**(1), 203–215 (2007)
6. D. Chae, Euler's equations and the maximum principle. Math. Ann. **361**, 51–66 (2015)
7. D. Chae, Unique continuation type theorem for the self-similar Euler equations. Adv. Math. **283**, 143–154 (2015)
8. D. Chae, On the generalized self-similar singularities for the Euler and the Navier-Stokes equations. J. Funct. Anal. **258**(9), 2865–2883 (2010)
9. D. Chae, Nonexistence of asymptotically self-similar singularities in the Euler and the Navier-Stokes equations. Math. Ann. **338**(2), 435–449 (2007)
10. D. Chae, On the finite-time singularities of the 3D incompressible Euler equations. Commun. Pure Appl. Math. **LX**, 0597–0617 (2007)
11. D. Chae, Global regularity for the 2D Boussinesq equations with partial viscosity terms. Adv. Math. **203**(2), 497–513 (2006)
12. D. Chae, Local existence and blow-up criterion for the Euler equations in the Besov spaces. Asymptot. Anal. **38**(3–4), 339–358 (2004)
13. D. Chae, On the well-posedness of the Euler equations in the Triebel-Lizorkin spaces. Commun. Pure Appl. Math. **55**(5), 654–678 (2002)
14. D. Chae, P. Constantin, On a Type I singularity condition in terms of the pressure for the Euler eq in $\mathbb{R}^3$, Int. Math. Res. Notices (to appear)
15. D. Chae, P. Constantin, Remarks on type I blow-up for the 3D Euler equations and the 2D Boussinesq equations, submitted
16. D. Chae, H.-S. Nam, Local existence and blow-up criterion for the Boussinesq equations. Proc. Roy. Soc. Edinburgh Sect. A **127**(5), 935–94 (1997)
17. D. Chae, S.-K. Kim, H.S. Nam, Local existence and blow-up criterion of Hölder continuous solutions of the Boussinesq equations. Nagoya Math. J. **155**, 55–80 (1999)
18. D. Chae, R. Shvydkoy, On formation of a locally self-similar collapse in the incompressible Euler equations. Arch. Ration. Mech. Anal. **209**(3), 999–1017 (2013)
19. D. Chae, J. Wolf, Localized non blow-up criterion of the Beale-Kato-Majda type for the 3D Euler equations. Math. Ann. (to appear)
20. D. Chae, J. Wolf, The Euler equations in a critical case of the generalized Campanato space. Ann. Inst. H. Poincare Anal. Non Linéaire **38**(2), 201–241 (2021)
21. D. Chae, J. Wolf, *Local non blow-up condition of $C^{1,\alpha}$ solutions of the 3D incompressible Euler equations,* (in preparation)
22. D. Chae, J. Wolf, Energy concentrations and Type I blow-up for the 3D Euler equations. Comm. Math. Phys. **376**(2), 1627–1669 (2020)
23. D. Chae, J. Wolf, Removing type II singularities off the axis for the three dimensional axisymmetric Euler equations. Arch. Ration. Mech. Anal. **23**(3), 1041–1089 (2019)
24. D. Chae, J. Wolf, On the local Type I conditions for the 3D Euler equations. Arch. Rat. Mech. Anal. **2**, 641–663 (2018)
25. D. Chae, J. Wolf, On the regularity of solutions to the 2D Boussinesq equations satisfying type I conditions. J. Nonlinear Sci. **29**(2), 643–654 (2019)
26. J.-Y. Chemin, *Perfect Incompressible Fluids*, Oxford Lecture Series in Mathematics and its Applications, vol. 14. (Oxford Univ. Press, 1998)

27. J. Chen, T.-Y. Hou, Finite time blow-up of 2D Boussinesq and 3D Euler equations with velocity and boundary. Arch. Ration. Mech. Anal. **383**(3), 1559–1667 (2021)
28. Q. Chen, C. Miao, Z. Zhang, On the well-posedness of the ideal MHD equations in the Triebel-Lizorkin spaces. Arch. Ration. Mech. Anal. **195**(2), 561–578 (2010)
29. A.J. Chorin, J.E. Marsden, *A Mathematical Introduction to Fluid Mechanics* (Springer, 1990)
30. P. Constantin, *Analysis of Hydrodynamic Models*, CBMS-NSF Regional Conf. Ser. in Applied Mathematics, vol. 90. (SIAM, 2016)
31. P. Constantin, On the Euler equations of incompressible fluids. Bull. Amer. Math. Soc. **44**(4), 603–621 (2007)
32. P. Constantin, Geometric statistics in turbulence. SIAM Rev. **36**, 73–98 (1994)
33. P. Constantin, C. Fefferman, A. Majda, Geometric constraints on potential singularity formulation in the 3-D Euler equations. Comm. P.D.E. **21**(3–4), 559–571 (1996)
34. D. Cordoba, C. Fefferman, R. De La LLave, On squirt singularities in hydrodynamics. SIAM J. Math. Anal. **36**(1), 204–213 (2004)
35. J. Deng, T.Y. Hou, X. Yu, Improved geometric conditions for non-blow-up of the 3D incompressible Euler equations. Comm. P.D.E. **31**(1–3), 293–306 (2006)
36. T. Elgindi, I-J. Jeong, Finite-time singularity formation for strong solutions to the Boussinesq system. Ann. PDE **6**(1), Paper No. 5, 50 pp. (2020)
37. T. Elgindi, I-J. Jeong, Finite-time singularity formation for strong solutions to the axi-symmetric 3D Euler equations. Ann. PDE **5**(2), Paper No. 16, 51 pp. (2019)
38. L. Euler, Principes généraux du mouvement des fluides. Mémoires de l'académie des sciences de Berlin **11**, 274–315 (1755)
39. F. John, L. Nirenberg, On functions of bounded mean oscillation. Commun. Pure Appl. Math. **343**, 415–426 (1983)
40. T. Kato, Nonstationary flows of viscous and ideal fluids in $\mathbb{R}^3$. J. Funct. Anal. **9**, 296–305 (1972)
41. T. Kato, G. Ponce, Commutator estimates and the Euler and Navier-Stokes equations. Commun. Pure Appl. Math. **41**(7), 891–907 (1988)
42. R.M. Kerr, Evidence for a singularity of the three-dimensional, incompressible Euler equations. Phys. Fluids, A **5**(7), 1725–1746 (1993)
43. S. Klainerman, A. Majda, Singular limits of quasilinear hyperbolic systems with large parameters and the incompressible limit of compressible fluids. Commun. Pure Appl. Math. **34**, 481–524 (1981)
44. H. Kozono, Y. Taniuchi, Bilinear estimates in BMO and the Navier-Stokes equations. Math. Z. **235**(1), 173–194 (2000)
45. H. Kozono, Y. Taniuchi, Limiting case of the Sobolev inequality in BMO, with applications to the Euler equations. Comm. Math. Phys. **214**, 191–200 (2000)
46. P.-L. Lions, *Mathematical Topics in Fluid Mechanics, Vol 1. Incompressible Models*, Oxford Lecture Series in Mathematics and its Applications, vol. 13. (Oxford Univ. Press, 1996)
47. G. Luo, T.-Y. Hou, Formation of finite-time singularities in the 3D axisymmetric Euler equations: a numerics guided study. SIAM Rev. **61**(4), 793–835 (2019)
48. G. Luo, T.-Y. Hou, Toward the finite-time blowup of the 3D axisymmetric Euler equations: a numerical investigation. Multiscale Model. Simul. **12**(4), 1722–1776 (2014)
49. A. Majda, A. Bertozzi, *Vorticity and Incompressible Flow* (Cambridge Univ. Press, 2002)
50. A. Majda, *Introduction to PDEs and Waves for the Atmsphere and Ocean*, Courant Lecture Note Series N. 9 (AMS, 2003)
51. H.K. Moffatt, Some remarks on topological fluid mechanics, in *An Introduction to the Geometry and Topology of Fluid Flows*. ed. by R.L. Ricca (Kluwer Academic Publishers, Dordrecht, The Netherlands, 2001), pp. 3–10
52. R. Shvydkoy, A study of energy concentration and drain in incompressible fluids. Nonlinearity **26**, 425–438 (2013)
53. E.M. Stein, *Harmonic Analysis: Real-variable Methods, Orthogonality, and Oscillatory Integrals* (Princeton University Press, 1993)
54. T. Tao, Finite time blowup for Lagrangian modifications of the three-dimensional Euler equation. Ann. PDE **2**(9), 1–79 (2016)

Singularity Models in the Three-Dimensional Ricci Flow

Simon Brendle

Abstract The Ricci flow is a natural evolution equation for Riemannian metrics on a given manifold. The main goal is to understand singularity formation. In his spectacular 2002 breakthrough, Perelman achieved a qualitative understanding of singularity formation in dimension 3. More precisely, Perelman showed that every finite-time singularity to the Ricci flow in dimension 3 is modeled on an ancient κ-solution. Moreover, Perelman proved a structure theorem for ancient κ-solutions in dimension 3. In this survey, we discuss recent developments which have led to a complete classification of all the singularity models in dimension 3. Moreover, we give an alternative proof of the classification of noncollapsed steady gradient Ricci solitons in dimension 3 (originally proved by the author in 2012).

Keywords Ricci flow · Ricci soliton · Ancient solution

1 Background on the Ricci Flow

Geometric evolution equations play an important role in differential geometry. The most important such evolution equation is the Ricci flow for Riemannian metrics which was introduced by Hamilton [21].

Definition 1.1 (*R. Hamilton* [21]) Let $g(t)$ be a one-parameter family of Riemannian metrics on a manifold M. We say that the metrics $g(t)$ evolve by the Ricci flow if

$$\frac{\partial}{\partial t} g(t) = -2 \operatorname{Ric}_{g(t)}. \tag{1}$$

The author was supported by the National Science Foundation under grants DMS-1806190 and DMS-2103573 and by the Simons Foundation.

S. Brendle (✉)
Department of Mathematics, Columbia University, 2990 Broadway, New York, NY 10027, USA
e-mail: simon.brendle@columbia.edu

In his paper [21], Hamilton established short time existence and uniqueness for the Ricci flow.

Theorem 1.2 (R. Hamilton [21]; D. DeTurck [18]) *Let g_0 be a Riemannian metric on a compact manifold M. Then there exists a unique solution $g(t)$, $t \in [0, T)$, to the Ricci flow with initial metric $g(0) = g_0$. Here, T is a positive real number which depends on the initial data.*

The main difficulty in proving Theorem 1.2 is that the Ricci flow is weakly, but not strictly, parabolic. This is due to the fact that the Ricci flow is invariant under the diffeomorphism group of M. This problem can be overcome using DeTurck's trick [18]. In the following, we sketch the argument (see [6] or [30] for details). Let us fix a compact manifold M and a smooth one-parameter family of background metrics $h(t)$. The choice of the background metrics $h(t)$ is not important. In particular, we can choose the background metrics $h(t)$ to be independent of t. For each t, we denote by $\Delta_{g(t),h(t)}$ the Laplacian of a map from $(M, g(t))$ to $(M, h(t))$ (see [6], Definition 2.2). With this understood, we can define Ricci-DeTurck flow as follows:

Definition 1.3 Let $\tilde{g}(t)$ be a one-parameter family of metrics on M. We say that the metrics $\tilde{g}(t)$ evolve by the Ricci-DeTurck flow if

$$\frac{\partial}{\partial t}\tilde{g}(t) = -2 \ \mathrm{Ric}_{\tilde{g}(t)} - \mathcal{L}_{\xi_t}(\tilde{g}(t)),$$

where the vector field ξ_t is definrd by $\xi_t := \Delta_{\tilde{g}(t),h(t)}$ id.

The evolution of the metric under the Ricci-DeTurck flow can be written in the form

$$\frac{\partial}{\partial t}\tilde{g}_{ij} = \tilde{g}^{kl} \ \partial_k \partial_l \tilde{g}_{ij} + \text{ lower order terms.}$$

Therefore, the Ricci-DeTurck flow is strictly parabolic, and admits a unique solution on a short time interval.

In the next step, we show that the Ricci flow is equivalent to the Ricci-DeTurck flow in the sense that whenever we have a solution to one equation, we can convert it into a solution of the other.

To explain this, suppose first that we are given a solution $\tilde{g}(t)$ of the Ricci-DeTurck flow. Our goal is to produce a solution $g(t)$ of the Ricci flow. As above, we define $\xi_t := \Delta_{\tilde{g}(t),h(t)}$ id. We define a one-parameter family of diffeomorphisms φ_t by $\frac{\partial}{\partial t}\varphi_t(p) = \xi_t|_{\varphi_t(p)}$ and $\varphi_0(p) = p$. Moreover, we define a one-parameter family of metrics $g(t)$ by $g(t) = \varphi_t^*(\tilde{g}(t))$. Then $g(t)$ is a solution of the Ricci flow.

Conversely, suppose that we are given a solution $g(t)$ of the Ricci flow. Our goal is to produce a solution $\tilde{g}(t)$ of the Ricci-DeTurck flow. To that end, we solve the harmonic map heat flow $\frac{\partial}{\partial t}\varphi_t = \Delta_{g(t),h(t)}\varphi_t$ with initial condition $\varphi_0 = $ id. Moreover, we define a one-parameter family of metrics $\tilde{g}(t)$ by $\varphi_t^*(\tilde{g}(t)) = g(t)$. Then $\tilde{g}(t)$ is a solution of the Ricci-DeTurck flow. This shows that the Ricci flow is equivalent to the Ricci-DeTurck flow.

A solution to the Ricci flow on a compact manifold can either be continued for all time, or else the curvature must blow up in finite time:

Theorem 1.4 (R. Hamilton [21]) *Let g_0 be a Riemannian metric on a compact manifold M. Let $g(t)$, $t \in [0, T)$, denote the unique maximal solution to the Ricci flow with initial metric $g(0) = g_0$. If $T < \infty$, then the curvature of $g(t)$ is unbounded as $t \to T$.*

A central problem is to understand the formation of singularities under the Ricci flow. To that end, it is often useful to consider a special class of solutions which move in a self-similar fashion. These are referred to as Ricci solitons:

Definition 1.5 Let (M, g) be a Riemannian manifold, and let f be a scalar function on M. We say that (M, g, f) is a steady gradient Ricci soliton if $\mathrm{Ric} = D^2 f$. We say that (M, g, f) is a shrinking gradient Ricci soliton if $\mathrm{Ric} = D^2 f + \mu g$ for some constant $\mu > 0$. We say that (M, g, f) is an expanding gradient Ricci soliton if $\mathrm{Ric} = D^2 f + \mu g$ for some constant $\mu < 0$.

We next discuss the global behavior of the Ricci flow in dimension 2. Hamilton [22] and Chow [15] showed that, for every initial metric on S^2, the Ricci flow shrinks to a point and becomes round after rescaling:

Theorem 1.6 (R. Hamilton [22]; B. Chow [15]) *Let g_0 be a Riemannian metric on S^2. Let $g(t)$, $t \in [0, T)$, denote the unique maximal solution to the Ricci flow with initial metric $g(0) = g_0$. Then $T < \infty$. Moreover, as $t \to T$, the rescaled metrics $\frac{1}{2(T-t)} g(t)$ converge in C^∞ to a metric with constant Gaussian curvature 1.*

Theorem 1.6 was first proved by Hamilton [22] under the additional assumption that the initial metric g_0 has positive scalar curvature. This condition was later removed by Chow [15]. In the following, we sketch the main ideas in Hamilton's proof. Full details can be found in [22] or [6], Sect. 4. Given a metric g on S^2 with positive scalar curvature, Hamilton defines the entropy $\mathcal{E}(g)$ by

$$\mathcal{E}(g) = \int_{S^2} R \, \log\left(\frac{AR}{8\pi}\right) d\mu, \tag{2}$$

where A denotes the area of (S^2, g). The functional $\mathcal{E}(g)$ is invariant under scaling. By the Gauss-Bonnet theorem, $\int_{S^2} R \, d\mu = 8\pi$. Hence, it follows from Jensen's inequality that $\mathcal{E}(g)$ is nonnegative. Moreover, $\mathcal{E}(g)$ is strictly positive unless the scalar curvature of (S^2, g) is constant.

Hamilton's key insight is that the functional $\mathcal{E}(g)$ is monotone decreasing under the Ricci flow. From this, Hamilton deduced that the product AR is uniformly bounded under the evolution. This implies that the flow converges to a shrinking gradient Ricci soliton, up to scaling. Finally, Hamilton showed that every shrinking gradient Ricci soliton on S^2 must have constant scalar curvature. This completes our discussion of Theorem 1.6.

In the three-dimensional case, Hamilton [21] showed that an initial metric with positive Ricci curvature shrinks to a point in finite time and becomes round after rescaling.

Theorem 1.7 (R. Hamilton [21]) *Let g_0 be a Riemannian metric on a three-manifold M with positive Ricci curvature. Let $g(t)$, $t \in [0, T)$, denote the unique maximal solution to the Ricci flow with initial metric $g(0) = g_0$. Then $T < \infty$. Moreover, as $t \to T$, the rescaled metrics $\frac{1}{4(T-t)} g(t)$ converge in C^∞ to a metric with constant sectional curvature 1.*

The proof of Theorem 1.7 is based on a pinching estimate for the eigenvalues of the Ricci tensor. To explain this, let $\lambda_1 \le \lambda_2 \le \lambda_3$ denote the eigenvalues of the tensor $R \, g_{ij} - 2 \, \mathrm{Ric}_{ij}$. With this understood, the scalar curvature is given by $\lambda_1 + \lambda_2 + \lambda_3$, and the eigenvalues of the Ricci tensor are given by $\frac{1}{2}(\lambda_2 + \lambda_3)$, $\frac{1}{2}(\lambda_3 + \lambda_1)$, $\frac{1}{2}(\lambda_1 + \lambda_2)$. In particular, the positivity of the Ricci tensor is equivalent to the inequality $\lambda_1 + \lambda_2 > 0$. Hamilton proved that

$$\frac{\partial}{\partial t}\lambda_1 \ge \Delta\lambda_1 + \lambda_1^2 + \lambda_2\lambda_3 \tag{3}$$

and

$$\frac{\partial}{\partial t}\lambda_3 \le \Delta\lambda_3 + \lambda_3^2 + \lambda_1\lambda_2, \tag{4}$$

where both inequalities are understood in the barrier sense. In the special case when the initial metric has positive Ricci curvature, Hamilton proved a pinching estimate of the form $\lambda_3 - \lambda_1 \le C \, (\lambda_1 + \lambda_2)^{1-\delta}$, where δ is a small positive constant depending on the initial data and C is a large constant depending on the initial data. The proof of this estimate relies on the maximum principle.

Theorem 1.7 has opened up two major lines of research. On the one hand, it is of interest to prove similar convergence theorems in higher dimensions, under suitable assumptions on the curvature. This direction led to the proof of the Differentiable Sphere Theorem (see [5, 12]). On the other hand, it is important to understand the behavior of the Ricci flow in dimension 3 for arbitrary initial metrics. In this case, the flow will develop more complicated types of singularities, including so-called neck-pinch singularities. In a series of breakthroughs, Perelman [27, 28] achieved a qualitative understanding of singularity formation in dimension 3. This is sufficient for topological conclusions, such as the Poincaré conjecture.

In this survey, we will focus on issues related to singularity formation in dimension 3. In Sect. 2, we will review the concept of an ancient solution, and explain its relevance for the analysis of singularities. We next discuss Perelman's noncollapsing theorem. Moreover, we describe examples of ancient solutions to the Ricci flow in low dimensions. In Sect. 3, we discuss the classification of ancient solutions in dimension 2. In Sect. 4, we review results due to Perelman [27] concerning the structure of ancient κ-solutions in dimension 3. These are ancient solutions which have bounded and nonnegative curvature and satisfy a noncollapsing condition. In

Sect. 5, we discuss the classification of ancient κ-solutions in dimension 3. In Sect. 6, we describe a quantitative version of the fact that the Ricci flow preserves symmetry. In Sect. 7, we discuss the Neck Improvement Theorem from [8]. This theorem asserts that a neck becomes more symmetric under the evolution. Finally, in Sects. 8 and 9, we give an alternative proof of the classification of noncollapsed steady gradient Ricci solitons in dimension 3. This result was originally proved in [7]; the proof given here relies on the Neck Improvement Theorem from [8].

2 Ancient Solutions and Noncollapsing

The notion of an ancient solution plays a fundamental role in understanding the formation of singularities in the Ricci flow. This concept was introduced by Hamilton [23].

Definition 2.1 An ancient solution to the Ricci flow is a solution which is defined on the time interval $(-\infty, T]$ for some T.

The concept of an ancient solution to a parabolic PDE is analogous to the concept of an entire solution to an elliptic PDE.

Ancient solutions typically arise as blow-up limits at a singularity. In the Ricci flow, we are specifically interested in ancient solutions which satisfy a noncollapsing condition.

Definition 2.2 (*G. Perelman* [27]) An ancient solution to the Ricci flow in dimension n is said to be κ-noncollapsed if $\mathrm{vol}_{g(t)}(B_{g(t)}(p, r)) \geq \kappa r^n$ whenever $\sup_{x \in B_{g(t)}(p,r)} R(x, t) \leq r^{-2}$.

Definition 2.2 is motivated by Perelman's noncollapsing theorem for the Ricci flow:

Theorem 2.3 (G. Perelman [27], Sect. 4) *Let M be a compact manifold of dimension n, and let $g(t)$, $t \in [0, T)$, be a solution to the Ricci flow, where $T < \infty$. Consider a sequence of times $t_j \to T$ and a bounded sequence of radii r_j. Finally, let p_j be a sequence of points in M such that*

$$r_j^2 \sup_{x \in B_{g(t_j)}(p_j, r_j)} R(x, t_j) < \infty.$$

Then

$$\liminf_{j \to \infty} r_j^{-n} \, \mathrm{vol}_{g(t_j)}(B_{g(t_j)}(p_j, r_j)) > 0.$$

Theorem 2.3 is a consequence of Perelman's monotonicity formula for the $\mathcal{W}$-functional. In particular, Theorem 2.3 implies that every blow-up limit of the Ricci flow at a finite-time singularity must be κ-noncollapsed.

In dimension 3, the Hamilton-Ivey estimate gives a lower bound for the sectional curvature in terms of the scalar curvature:

Theorem 2.4 (R. Hamilton [23]; T. Ivey [25]) *Let $g(t)$, $t \in [0, T)$, be a solution to the Ricci flow on a compact three-manifold M. Let λ_1 denote the smallest eigenvalue of the tensor $R g_{ij} - 2 Ric_{ij}$. Then λ_1 satisfies a pointwise inequality of the form $\lambda_1 \geq -f(R)$, where the function f satisfies $\lim_{s \to \infty} \frac{f(s)}{s} = 0$.*

Theorem 2.4 implies that every blow-up limit of the Ricci flow in dimension 3 must have nonnegative sectional curvature. The proof of Theorem 2.4 relies on the maximum principle together with the evolution equation for the Ricci tensor.

This motivates the following definition:

Definition 2.5 An ancient κ-solution to the Ricci flow in dimension $n \in \{2, 3\}$ is a complete, non-flat, κ-noncollapsed ancient solution with bounded and nonnegative curvature.

The notion of an ancient κ-solutions plays a key role in Perelman's theory. In particular, Perelman showed that, if a solution to the Ricci flow in dimension 3 forms a singularity in finite time, then the high curvature regions can be approximated by ancient κ-solutions (see [27], Sect. 12).

In the remainder of this section, we describe some of the known examples of ancient solutions to the Ricci flow in dimension 2 and 3.

Example 2.6 Let g_{S^2} denote the standard metric on S^2. Let us define a family of metrics $g(t)$ on S^2 by $g(t) = (-2t) g_{S^2}$ for $t \in (-\infty, 0)$. This is an ancient solution to the Ricci flow which shrinks homothetically. It is κ-noncollapsed.

Example 2.7 Let us define a one-parameter family of conformal metrics on $\mathbb{R}^2$ by

$$g_{ij}(t) = \frac{4}{e^t + |x|^2} \delta_{ij}$$

for $t \in (-\infty, \infty)$. This gives a rotationally symmetric solution to the Ricci flow on $\mathbb{R}^2$, which moves by dieomorphisms. It is referred to as the cigar soliton. The cigar soliton has positive curvature and opens up like a cylinder near infinity. The cigar soliton fails to be κ-noncollapsed.

Example 2.8 Let us define a one-parameter family of conformal metrics on $\mathbb{R}^2$ by

$$g_{ij}(t) = \frac{8 \sinh(-t)}{1 + 2 \cosh(-t) |x|^2 + |x|^4} \delta_{ij}$$

for $t \in (-\infty, 0)$. For each $t \in (-\infty, 0)$, $g(t)$ extends to a smooth metric on S^2. This gives a rotationally symmetric solution to the Ricci flow on S^2. This is referred to as the King-Rosenau solution (cf. [26, 29]). The King-Rosenau solution is an ancient solution to the Ricci flow with positive curvature. The King-Rosenau solution fails to be κ-noncollapsed.

Example 2.9 Let g_{S^3} denote the standard metric on S^3. Let us define a family of metrics $g(t)$ on S^2 by $g(t) = (-4t) g_{S^3}$ for $t \in (-\infty, 0)$. This is an ancient solution to the Ricci flow which shrinks homothetically. It is κ-noncollapsed.

Example 2.10 Let again g_{S^2} denote the standard metric on S^2. Let us define a family of metrics $g(t)$ on $S^2 \times \mathbb{R}$ by $g(t) = (-2t) g_{S^2} + dz \otimes dz$ for $t \in (-\infty, 0)$. This is an ancient solution to the Ricci flow. It is κ-noncollapsed.

Example 2.11 Robert Bryant [13] has constructed a steady gradient Ricci soliton in dimension 3 which is rotationally symmetric. This can be viewed as the three-dimensional analogue of the cigar soliton. The Bryant soliton has positive sectional curvature and opens up like a paraboloid near infinity. Unlike the cigar soliton, the Bryant soliton is κ-noncollapsed.

Example 2.12 Perelman has constructed an ancient solution to the Ricci flow on S^3 which is rotationally symmetric. This can be viewed as the three-dimensional analogue of the King-Rosenau solution. Perelman's ancient solution has positive sectional curvature. Unlike the King-Rosenau solution, Perelman's ancient solution is κ-noncollapsed.

The asymptotics of Perelman's ancient solution are by now well understood; see [2].

3 Classification of Ancient Solutions in Dimension 2

In this section, we discuss the main classification results for ancient solution in dimension 2. In [27], Perelman gave a classification of ancient κ-solutions in dimension 2:

Theorem 3.1 (G. Perelman [27], Sect. 11) *Let $(M, g(t))$ be an ancient κ-solution in dimension 2. Then $(M, g(t))$ is isometric to a family of shrinking spheres, or a $\mathbb{Z}_2$-quotient thereof.*

Let us sketch Perelman's proof of Theorem 3.1. Suppose that $(M, g(t))$ is an ancient κ-solution in dimension 2. After passing to a double cover if necessary, we may assume that M is orientable. By Proposition 11.2 in [27], we can find a sequence of times $t_j \to -\infty$ and a sequence of points $p_j \in M$ with the following property: if we dilate the manifold $(M, g(t_j))$ around the point p_j by the factor $(-t_j)^{-\frac{1}{2}}$, then the rescaled manifolds converge in the Cheeger-Gromov sense to a non-flat shrinking gradient Ricci soliton. Using Hamilton's classification of shrinking gradient Ricci solitons in dimension 2 (see [22]), we conclude that the limiting manifold must be a round sphere. Since the limiting manifold is diffeomorphic to S^2, it follows that M is diffeomorphic to S^2.

We next consider Hamilton's entropy functional defined in (2). Since the manifolds $(M, g(t_j))$ converge to a round sphere after rescaling, we know that $\mathcal{E}(g(t_j)) \to 0$

as $j \to \infty$. Moreover, it follows from Hamilton's work [22] that the function $t \mapsto \mathcal{E}(g(t))$ is monotone decreasing. Consequently, $\mathcal{E}(g(t)) \leq 0$ for each t. On the other hand, Jensen's inequality implies that $\mathcal{E}(g(t))$ is strictly positive unless $(M, g(t))$ has constant scalar curvature. Putting these facts together, we conclude that the scalar curvature of $(M, g(t))$ is constant for each t. This completes our sketch of the proof of Theorem 3.1.

Daskalopoulos, Hamilton, and Šešum were able to classify compact ancient solutions in dimension 2 without noncollapsing assumptions.

Theorem 3.2 (P. Daskalopoulos, R. Hamilton, N. Šešum [17]) *Let $(M, g(t))$ be a compact, non-flat ancient solution to the Ricci flow in dimension 2. Then, up to parabolic rescaling, translation in time, and diffeomorphisms, $(M, g(t))$ coincides with the family of shrinking spheres, the King-Rosenau solution, or a $\mathbb{Z}_2$-quotient of these.*

One of the main ideas behind Theorem 3.2 is to find a quantity to which the maximum principle can be applied and which vanishes on the King-Rosenau solution. To explain this, suppose that $(M, g(t))$ is a compact, non-flat ancient solution to the Ricci flow in dimension 2. After passing to a double cover if necessary, we may assume that M is orientable. Using the maximum principle, it is easy to see that $(M, g(t))$ has nonnegative scalar curvature for each t. Moreover, the strict maximum principle implies that the scalar curvature of $(M, g(t))$ is strictly positive. Since M is compact and orientable, it follows that $M = S^2$. By the uniformization theorem, we may assume that the metrics $g(t)$ are conformal to the standard metric on S^2.

Using the stereographic projection, we can identify $\mathbb{R}^2$ with the complement of the north pole in S^2. Thus, we obtain a family of conformal metrics on $\mathbb{R}^2$ which evolve by the Ricci flow. We may write the evolving metric in the form $v^{-1} \delta_{ij}$, where v satisfies the parabolic PDE

$$\frac{\partial}{\partial t} v = v \, \Delta v - |\nabla v|^2$$

on $\mathbb{R}^2$, where ∇v and Δv denote the gradient and Laplacian of v with respect to the Euclidean metric on $\mathbb{R}^2$. Daskalopoulos, Hamilton, and Šešum consider the quantity

$$Q = v \, \frac{\partial^3 v}{\partial z^3} \, \frac{\partial^3 v}{\partial \bar{z}^3},$$

where $\frac{\partial}{\partial z} = \frac{1}{2} \left(\frac{\partial}{\partial x_1} - i \frac{\partial}{\partial x_2} \right)$ and $\frac{\partial}{\partial \bar{z}} = \frac{1}{2} \left(\frac{\partial}{\partial x_1} + i \frac{\partial}{\partial x_2} \right)$ denote the usual Wirtinger derivatives. A straightforward calculation shows that the quantity Q is invariant under Möbius transformations. Moreover, Q satisfies the evolution equation

$$\frac{\partial}{\partial t}Q = v\,\Delta Q - 16v^{-1}\left(v\,\frac{\partial^2 v}{\partial z \partial \bar{z}} - \frac{\partial v}{\partial z}\frac{\partial v}{\partial \bar{z}}\right)Q$$
$$- 4\left(v\,\frac{\partial^4 v}{\partial z^4} + 2\,\frac{\partial v}{\partial z}\frac{\partial^3 v}{\partial z^3}\right)\left(v\,\frac{\partial^4 v}{\partial \bar{z}^4} + 2\,\frac{\partial v}{\partial \bar{z}}\frac{\partial^3 v}{\partial \bar{z}^3}\right)$$
$$- 4\left(v\,\frac{\partial^4 v}{\partial z^3 \partial \bar{z}} - \frac{\partial v}{\partial \bar{z}}\frac{\partial^3 v}{\partial z^3}\right)\left(v\,\frac{\partial^4 v}{\partial \bar{z}^3 \partial z} - \frac{\partial v}{\partial z}\frac{\partial^3 v}{\partial \bar{z}^3}\right).$$

The scalar curvature of the conformal metric $v^{-1}\,\delta_{ij}$ can be written in the form

$$R = v^{-1}\left(v\,\Delta v - |\nabla v|^2\right) = 4v^{-1}\left(v\,\frac{\partial^2 v}{\partial z \partial \bar{z}} - \frac{\partial v}{\partial z}\frac{\partial v}{\partial \bar{z}}\right).$$

This gives

$$\frac{\partial}{\partial t}Q \leq v\,\Delta Q - 4RQ$$

(compare [17], Sect. 5, or [16]). Here, ΔQ denotes the Laplacian of Q with respect to the Euclidean metric on $\mathbb{R}^2$. The term $v\,\Delta Q$ can be interpreted as the Laplacian of Q with respect to the evolving metric $v^{-1}\,\delta_{ij}$.

On the King-Rosenau solution, v is a quadratic polynomial in $|x|^2$, with coefficients that depend on t. In particular, $\frac{\partial^3 v}{\partial z^3}$ vanishes identically on the King-Rosenau solution. Therefore, Q vanishes identically on the King-Rosenau solution.

4 Structure of Ancient κ-Solutions in Dimension 3

In [27], Perelman proved several fundamental results concerning the structure of ancient κ-solutions in dimension 3. One of the central results is the following pointwise estimate for the covariant derivatives of the curvature tensor.

Theorem 4.1 (G. Perelman [27], Sect. 11) *Let $(M, g(t))$, $t \in (-\infty, 0]$, be an ancient κ-solution to the Ricci flow in dimension 3. Let m be a positive integer. Then the m-th order covariant derivatives of the curvature tensor satisfy the pointwise bound $|D^m\, Rm| \leq C\,R^{\frac{m+2}{2}}$, where C is a positive constant that depends only on m and κ.*

Another fundamental result in Perelman's work is the following longrange curvature estimate:

Theorem 4.2 (G. Perelman [27], Sect. 11) *Let $(M, g(t))$, $t \in (-\infty, 0]$, be an ancient κ-solution to the Ricci flow in dimension 3. Then there exists a function $\omega : [0, \infty) \to [0, \infty)$ (depending on κ) such that*

$$R(y, t) \leq R(x, t)\,\omega(R(x, t)\,d_{g(t)}(x, y)^2)$$

for all x, $y \in M$ and all $t \leq 0$.

Perelman's longrange curvature estimate is extremely useful, in that it allows Perelman to take limits of sequences of ancient κ-solutions. One important consequence is that the space of ancient κ-solutions is compact in the following sense:

Theorem 4.3 (G. Perelman [27], Sect. 11) *Let $(M^{(j)}, g^{(j)}(t))$, $t \in (-\infty, 0]$, be a sequence of ancient κ-solutions in dimension 3. Moreover, suppose that $p_j \in M^{(j)}$ is a sequence of points such that $R(p_j, 0) = 1$ for each j. Then, after passing to a subsequence if necessary, the flows $(M^{(j)}, g^{(j)}(t), p_j)$ converge in the Cheeger-Gromov sense to a limit $(M^\infty, g^\infty(t))$, and this limit is again an ancient κ-solution.*

Corollary 4.4 (G. Perelman [27], Sect. 11) *Let $(M, g(t))$, $t \in (-\infty, 0]$, be a non-compact ancient κ-solution to the Ricci flow in dimension 3 with positive sectional curvature. Let us fix a point p_0 in M. Let p_j be a sequence of points in M such that $d_{g(0)}(p_0, p_j) \to \infty$, and let $r_j^{-2} := R(p_j, 0)$. Let us dilate the flow around the point $(p_j, 0)$ by the factor r_j^{-1}. Then, after passing to a subsequence if necessary, the rescaled flows converge in the Cheeger-Gromov sense to a family of shrinking cylinders.*

Let us sketch how Corollary 4.4 follows from Theorem 4.3. The longrange curvature estimate gives

$$R(p_0, 0) \leq R(p_j, 0)\, \omega(R(p_j, 0)\, d_{g(0)}(p_0, p_j)^2)$$

for each j. This implies

$$R(p_j, 0)\, d_{g(0)}(p_0, p_j)^2 \to \infty$$

as $j \to \infty$. In other words, $r_j^{-1} d_{g(0)}(p_0, p_j) \to \infty$ as $j \to \infty$. We next consider the rescaled metrics $g^{(j)}(t) := r_j^{-2} g(r_j^2 t)$ for $t \in (-\infty, 0]$. By Theorem 4.3, the flows $(M, g^{(j)}(t), p_j)$ converge in the Cheeger-Gromov sense to an ancient κ-solution $(M^\infty, g^\infty(t))$. Since $r_j^{-1} d_{g(0)}(p_0, p_j) \to \infty$, the limiting flow $(M^\infty, g^\infty(t))$ must split off a line. Using Perelman's classification of ancient κ-solutions in dimension 2 (see Theorem 3.1), it follows that the limiting flow $(M^\infty, g^\infty(t))$ must be a family of shrinking cylinders or a quotient thereof. On the other hand, M is diffeomorphic to $\mathbb{R}^3$ by the soul theorem. In particular, M does not contain an embedded $\mathbb{RP}^2$. This implies that $(M^\infty, g^\infty(t))$ cannot be a non-trivial quotient of the cylinder. This completes the sketch of the proof of Corollary 4.4.

Definition 4.5 Let $(M, g(t))$ be a solution to the Ricci flow in dimension 3, and let $(\bar{x}, \bar{t})$ be a point in space-time with $R(\bar{x}, \bar{t}) = r^{-2}$. We say that $(\bar{x}, \bar{t})$ lies at the center of an evolving ε-neck if, after rescaling by the factor r^{-1}, the parabolic neighborhood $B_{g(\bar{t})}(\bar{x}, \varepsilon^{-1} r) \times [\bar{t} - \varepsilon^{-1} r^2, \bar{t}]$ is ε-close in $C^{[\varepsilon^{-1}]}$ to a family of shrinking cylinders.

The notion of a neck was introduced in Hamilton's work [24]. In particular, Hamilton showed that a neck admits a canonical foliation by constant mean curvature (CMC) spheres.

Corollary 4.4 implies the following structure theorem for noncompact ancient κ-solutions:

Corollary 4.6 (G. Perelman [27], Sect. 11) *Let $(M, g(t))$, $t \in (-\infty, 0]$, be a noncompact ancient κ-solution to the Ricci flow in dimension 3 with positive sectional curvature. Moreover, let ε be a positive real number, and let M_ε denote the set of all points $x \in M$ with the property that $(x, 0)$ does not lie at the center of an evolving ε-neck. Then M_ε has finite diameter. Moreover, $\sup_{x \in M_\varepsilon} R(x, 0) \leq C(\kappa, \varepsilon) \inf_{x \in M_\varepsilon} R(x, 0)$ and $\sup_{x \in M_\varepsilon} R(x, 0) \leq C(\kappa, \varepsilon) \operatorname{diam}_{g(0)}(M_\varepsilon)^{-2}$.*

5 Classification of Ancient κ-Solutions in Dimension 3

We now turn to the classification of ancient κ-solutions in dimension 3. The first major step was the classification of noncollapsed steady gradient Ricci solitons in dimension 3.

Theorem 5.1 (S. Brendle [7]) *Let (M, g) be a three-dimensional complete steady gradient Ricci soliton which is non-flat and κ-noncollapsed. Then (M, g) is rotationally symmetric, and is therefore isometric to the Bryant soliton up to scaling.*

More recently, we classified all noncompact ancient κ-solutions in dimension 3:

Theorem 5.2 (S. Brendle [8]) *Assume that $(M, g(t))$ is a noncompact ancient κ-solution of dimension 3. Then either $(M, g(t))$ is isometric to a family of shrinking cylinders (or a quotient thereof), or $(M, g(t))$ is isometric to the Bryant soliton up to scaling.*

Theorem 5.2 confirms a conjecture of Perelman [27].

The proof of Theorem 5.2 consists of two main steps. In the first step, we classify noncompact ancient κ-solutions with rotational symmetry. To do that, we need precise asymptotic estimates for such solutions. In the second step, we show that every noncompact ancient κ-solution is rotationally symmetric. This second step uses the classification of steady gradient Ricci solitons in Theorem 5.1, as well as the classification of ancient κ-solutions with rotational symmetry. Another crucial ingredient is the Neck Improvement Theorem which asserts that a neck tends to get more symmetric as it evolves under the Ricci flow. We will discuss the Neck Improvement Theorem in Sect. 7 below.

The following theorem is the counterpart of Theorem 5.2 in the compact case:

Theorem 5.3 (S. Brendle, P. Daskalopoulos, N. Šešum [10]) *Assume that $(M, g(t))$ is a compact ancient κ-solution of dimension 3. Then, up to parabolic rescaling, translation in time, and diffeomorphisms, $(M, g(t))$ is either a family of shrinking spheres or Perelman's ancient solution or a quotient of these.*

The proof of Theorem 5.3 again requires two main steps. In the first step, we show that every compact ancient κ-solution is rotationally symmetric. This step uses the classification of noncompact ancient κ-solutions in Theorem 5.2, together with the Neck Improvement Theorem. In a second step, we classify compact ancient κ-solutions with rotational symmetry. To do that, we need to understand the asymptotic behavior of such solutions. These asymptotic estimates are established in [2].

Similar classification results exist for convex, noncollapsed ancient solutions to mean curvature flow in $\mathbb{R}^3$. We refer to [9] for the classification in the noncompact case, and to [3] for the classification in the compact case.

6   Preservation of Symmetry Under the Ricci Flow

Let $g(t)$, $t \in [0, T)$, be a solution to the Ricci flow on a compact manifold M. It follows from Hamilton's short time uniqueness theorem that the Ricci flow preserves symmetry. More precisely, every isometry of $(M, g(0))$ is also an isometry of $(M, g(t))$ for each $t \geq 0$. Consequently, every Killing vector field of $(M, g(0))$ is also a Killing vector field of $(M, g(t))$ for each $t \geq 0$.

In this section, we describe a quantitative version of this principle. We begin with a definition:

Definition 6.1 "Let h be a symmetric $(0; 2)$-tensor. The Lichnerowicz Laplacian of h is defined as

$$\Delta_L h_{ik} := \Delta h_{ik} + 2R_{ijkl}h^{jl} - \mathrm{Ric}_i^l h_{kl} - \mathrm{Ric}_k^l h_{il}. \tag{5}$$

The Lichnerowicz Laplacian arises naturally in the study of Einstein metrics (see [4], Eq. (1.180b)). It also comes up in connection with the evolution equation for the Ricci tensor under the Ricci flow. Indeed, if $(M, g(t))$ is a solution to the Ricci flow, then the Ricci tensor satisfies the evolution equation

$$\frac{\partial}{\partial t} \mathrm{Ric}_{g(t)} = \Delta_{L,g(t)} \mathrm{Ric}_{g(t)}$$

(see [21], Corollary 7.3).

The following results play a key role in our analysis:

Proposition 6.2 *Let (M, g) be a Riemannian manifold. Let V be a smooth vector field on M, and let $h := \mathscr{L}_V(g)$. Then*

$$\Delta V + Ric(V) = div\, h - \frac{1}{2}\nabla(\,tr\,h).$$

Proposition 6.3 (S. Brendle [8], Sect. 5) *Let $(M, g(t))$ be a solution of the Ricci flow. Let $V(t)$ be a time-dependent vector field such that*

$$\frac{\partial}{\partial t}V(t) = \Delta_{g(t)}V(t) + Ric_{g(t)}(V(t)), \tag{6}$$

and let $h(t) := \mathscr{L}_{V(t)}(g(t))$. Then the tensor $h(t)$ satisfies the evolution equation

$$\frac{\partial}{\partial t}h(t) = \Delta_{L,g(t)}h(t). \tag{7}$$

Proposition 6.3 has a natural geometric interpretation in terms of the linearized Ricci-DeTurck flow. To explain this, let us fix a solution $(M, g(t))$ of the Ricci flow. Suppose that φ_t is a one-parameter family of diffeomorphisms which solve the harmonic map heat flow with respect to the background metrics $g(t)$; that is,

$$\frac{\partial}{\partial t}\varphi_t = \Delta_{g(t),g(t)}\varphi_t. \tag{8}$$

Let us define a one-parameter family of metrics $\tilde{g}(t)$ by $\varphi_t^*(\tilde{g}(t)) = g(t)$. Then the metrics $\tilde{g}(t)$ solve the Ricci-DeTurck flow with respect to the background metrics $g(t)$. More precisely,

$$\frac{\partial}{\partial t}\tilde{g}(t) = -2 \ Ric_{\tilde{g}(t)} - \mathscr{L}_{\xi_t}(\tilde{g}(t)), \tag{9}$$

where $\xi_t = \Delta_{\tilde{g}(t),g(t)}$ id. Clearly, $\varphi_t := $ id is a solution of (8), and $\tilde{g}(t) := g(t)$ is a solution of (9).

We now linearize the Eqs. (8) and (9) around $\varphi_t = $ id and $\tilde{g}(t) = g(t)$, respectively. Linearizing the harmonic map heat flow (8) around the identity, we obtain the equation

$$\frac{\partial}{\partial t}V(t) = \Delta_{g(t)}V(t) + Ric_{g(t)}(V(t))$$

for a vector field V (see [19], p. 11). Linearizing the Ricci-DeTurck flow (9) around $g(t)$ leads to the parabolic Lichnerowicz equation

$$\frac{\partial}{\partial t}h(t) = \Delta_{L,g(t)}h(t).$$

This completes our discussion of Proposition 6.3.

On a steady gradient Ricci soliton, Proposition 6.3 takes the following form:

Corollary 6.4 (S. Brendle [7]) *Let (M, g, f) be a steady gradient Ricci soliton, and let $X := \nabla f$. Let V be a vector field satisfying*

$$\Delta V + D_X V = 0, \tag{10}$$

and let $h := \mathscr{L}_V(g)$. Then the tensor h satisfies the equation

$$\Delta_L h + \mathscr{L}_X(h) = 0. \tag{11}$$

Let us sketch how Corollary 6.4 follows from Theorem 6.3. On a steady gradient Ricci soliton, the time derivative $\frac{\partial}{\partial t}$ reduces to a Lie derivative $-\mathscr{L}_X$. More precisely, let (M, g, f) be a steady gradient Ricci soliton, let $X := \nabla f$, and let Φ_t denote the flow generated by the vector field $-X$. Suppose that V satisfies $\Delta V + D_X V = 0$, and let $h := \mathscr{L}_V(g)$. Using the identity $D_V X = \text{Ric}(V)$, we obtain $\Delta V + \mathscr{L}_X V + \text{Ric}(V) = 0$. Consequently, the vector fields $\Phi_t^*(V)$ satisfy the parabolic PDE (6) on the evolving background $(M, \Phi_t^*(g))$. By Theorem 6.3, the tensors $\Phi_t^*(h)$ satisfy the parabolic PDE (7) on the evolving background $(M, \Phi_t^*(g))$. This implies $\Delta_L h + \mathscr{L}_X(h) = 0$.

In order to apply Proposition 6.3 in practice, we need estimates for solutions of the parabolic Lichnerowicz equation. In dimension 3, this can be accomplished by applying the maximum principle to the quantity $\frac{|h|^2}{R^2}$:

Proposition 6.5 (G. Anderson, B. Chow [1]) *Let $(M, g(t))$ be a solution to the Ricci flow in dimension 3 with positive scalar curvature. Let h be a solution of the parabolic Lichnerowicz equation $\frac{\partial}{\partial t} h(t) = \Delta_{L,g(t)} h(t)$. Then*

$$\frac{\partial}{\partial t}\left(\frac{|h|^2}{R^2}\right) \leq \Delta\left(\frac{|h|^2}{R^2}\right) + \frac{2}{R}\left\langle \nabla R, \nabla\left(\frac{|h|^2}{R^2}\right)\right\rangle.$$

On a steady gradient Ricci soliton, Proposition 6.5 takes the following form:

Corollary 6.6 (G. Anderson, B. Chow [1]) *Let (M, g, f) be a steady gradient Ricci soliton in dimension 3 with positive scalar curvature, and let $X := \nabla f$. Let h be a solution of the equation $\Delta_L h + \mathscr{L}_X(h) = 0$. Then*

$$\Delta\left(\frac{|h|^2}{R^2}\right) + \left\langle X, \nabla\left(\frac{|h|^2}{R^2}\right)\right\rangle + \frac{2}{R}\left\langle \nabla R, \nabla\left(\frac{|h|^2}{R^2}\right)\right\rangle \geq 0.$$

7 Improvement of Symmetry on a Neck

In view of Proposition 6.3, it is important to understand the parabolic Lichnerowicz equation (7) on a Ricci flow background. As a starting point, we consider the special case when the background is given by a family of shrinking cylinders. To fix notation, we define a family of metrics $\bar{g}(t)$, $t \in (-\infty, 0)$ on $S^2 \times \mathbb{R}$ by

$$\bar{g}(t) = (-2t)\, g_{S^2} + dz \otimes dz, \quad t \in (-\infty, 0).$$

Clearly, the metrics $\bar{g}(t)$, $t \in (-\infty, 0)$, evolve by the Ricci flow.

Proposition 7.1 (S. Brendle [8], Sect. 6) *Let $(S^2 \times \mathbb{R}, \bar{g}(t))$ denote the family of shrinking cylinders. Let L be a large real number. Let $h(t)$ be a solution* *of the parabolic Lichnerowicz equation $\frac{\partial}{\partial t} h(t) = \Delta_{L, \bar{g}(t)} h(t)$ which is defined on $S^2 \times [-\frac{L}{2}, \frac{L}{2}]$ and for $t \in [-\frac{L}{2}, -1]$. Assume that $|h(t)|_{\bar{g}(t)} \leq 1$ for $t \in [-\frac{L}{2}, -\frac{L}{4}]$, and $|h(t)|_{\bar{g}(t)} \leq L^{10}$ for $t \in [-\frac{L}{4}, -1]$. Then we can find a rotationally invariant tensor of the form $\bar{\omega}(z, t) g_{S^2} + \bar{\beta}(z, t) dz \otimes dz$ and a scalar function $\psi : S^2 \to \mathbb{R}$ such that ψ lies in the span of the first spherical harmonics on S^2 and*

$$|h(t) - \bar{\omega}(z, t) g_{S^2} - \bar{\beta}(z, t) dz \otimes dz - (-t) \psi g_{S^2}|_{\bar{g}(t)} \leq C L^{-\frac{1}{2}}$$

on $S^2 \times [-1000, 1000]$ and for $t \in [-1000, -1]$. Here, C is a constant which does not depend on L.

Proposition 7.1 asserts that, given sufficient time to evolve, a solution of the parabolic Lichnerowicz equation can be approximated by a sum of a rotationally invariant tensor and a tensor of the form $(-t) \psi g_{S^2}$, where $\psi : S^2 \to \mathbb{R}$ lies in the span of the first spherical harmonics on S^2. The tensor $(-t) \psi g_{S^2}$ can be written as a Lie derivative of the metric along a vector field. To see this, let us define a vector field ξ on S^2 by $g_{S^2}(\xi, \cdot) = -\frac{1}{4} d\psi$. Since ψ lies in the span of the first spherical harmonics on S^2, we obtain $\mathscr{L}_\xi(g_{S^2}) = \frac{1}{2} \psi g_{S^2}$, and consequently $\mathscr{L}_\xi(\bar{g}(t)) = (-t) \psi g_{S^2}$.

To prove Proposition 7.1, we decompose the tensor $h(t)$ into components, and perform a mode decomposition in spherical harmonics. This leads to a system of linear heat equations in one space dimension.

Using Proposition 7.1, we can show that a neck becomes more symmetric as it evolves under the Ricci flow. To state this result, we need a quantitative notion of ε-symmetry:

Definition 7.2 (*S. Brendle* [8], *Sect. 8*) Let $(M, g(t))$ be a solution to the Ricci flow in dimension 3, and let $(\bar{x}, \bar{t})$ be a point in space-time with $R(\bar{x}, \bar{t}) = r^{-2}$. We assume that $(\bar{x}, \bar{t})$ lies at the center of an evolving ε_0-neck for some small positive number ε_0. We say that $(\bar{x}, \bar{t})$ is ε-symmetric if there exist smooth, time-independent vector fields $U^{(1)}, U^{(2)}, U^{(3)}$ which are defined on an open set containing $\bar{B}_{g(\bar{t})}(\bar{x}, 100r)$ and satisfy the following conditions:

- $\sup_{\bar{B}_{g(\bar{t})}(\bar{x}, 100r) \times [\bar{t} - 100r^2, \bar{t}]} \sum_{l=0}^{2} \sum_{a=1}^{3} r^{2l} |D^l (\mathscr{L}_{U^{(a)}}(g(t)))|^2 \leq \varepsilon^2$.
- If $t \in [\bar{t} - 100r^2, \bar{t}]$ and $\Sigma \subset \bar{B}_{g(\bar{t})}(\bar{x}, 100r)$ is a leaf of the CMC foliation of $(M, g(t))$, then $\sup_{\Sigma} \sum_{a=1}^{3} r^{-2} |\langle U^{(a)}, \nu \rangle|^2 \leq \varepsilon^2$, where ν denotes the unit normal vector to Σ in $(M, g(t))$.
- If $t \in [\bar{t} - 100r^2, \bar{t}]$ and $\Sigma \subset \bar{B}_{g(\bar{t})}(\bar{x}, 100r)$ is a leaf of the CMC foliation of $(M, g(t))$, then

$$\sum_{a,b=1}^{3} \left| \delta_{ab} - \operatorname{area}_{g(t)}(\Sigma)^{-2} \int_{\Sigma} \langle U^{(a)}, U^{(b)} \rangle_{g(t)} d\mu_{g(t)} \right|^2 \leq \varepsilon^2.$$

With this understood, we can now state the Neck Improvement Theorem from [8]:

Theorem 7.3 (S. Brendle [8], Sect. 8) *We can find a large constant L and small positive constant ε_1 such that the following holds. Let $(M, g(t))$ be a solution of the Ricci flow in dimension 3, and let (x_0, t_0) be a point in space-time which lies at the center of an evolving ε_1-neck and satisfies $R(x_0, t_0) = r^{-2}$. Moreover, we assume that every point in the parabolic neighborhood $B_{g(t_0)}(x_0, Lr) \times [t_0 - Lr^2, t_0)$ is ε-symmetric, where $\varepsilon \leq \varepsilon_1$. Then the point (x_0, t_0) is $\frac{\varepsilon}{2}$-symmetric.*

8 Asymptotic Behavior of Noncollapsed Steady Gradient Ricci Solitons in Dimension 3

Let (M, g, f) be a non-flat steady gradient Ricci soliton in dimension n, so that $\mathrm{Ric} = D^2 f$. For abbreviation, let $X := \nabla f$. Throughout this section, we fix an arbitrary point $p \in M$.

Lemma 8.1 *Given any point $\bar{x} \in M$, we can find a smooth function ρ which is defined in an open neighborhood of $\bar{x}$ and satisfies the following conditions:*

- *$\rho(x) \geq d(p, x)^2$ in an open neighborhood of the point $\bar{x}$.*
- *$\rho(\bar{x}) = d(p, \bar{x})^2$.*
- *$|\nabla \rho|^2 = 4\rho$ at the point $\bar{x}$.*
- *$\Delta \rho + \langle X, \nabla \rho \rangle \leq N_0 + N_1 \sqrt{\rho}$ at the point $\bar{x}$.*

Here, N_0 and N_1 are uniform constants which do not depend on $\bar{x}$.

Proof Let us fix a positive real number r_0 such that r_0 is strictly smaller than the injectivity radius at p.

We first consider the case $\bar{x} \in B(p, r_0)$. In this case, we define a smooth function $\rho : B(p, r_0) \to \mathbb{R}$ by $\rho(x) := d(p, x)^2$. It is easy to see that $|\nabla \rho|^2 = 4\rho$ at each point in $B(p, r_0)$. Moreover, at each point in $B(p, r_0)$, we have $\Delta \rho + \langle X, \nabla \rho \rangle \leq C$ for some uniform constant C.

In the next step, we consider the case $\bar{x} \in M \setminus B(p, r_0)$. Let $l := d(p, \bar{x}) \geq r_0$. Moreover, let $\gamma : [0, l] \to M$ be a unit-speed geodesic with $\gamma(0) = p$ and $\gamma(l) = \bar{x}$. Finally, let $\chi : [0, \infty) \to [0, \infty)$ be a smooth cutoff function such that $\chi = 0$ on the interval $[0, \frac{r_0}{2}]$ and $\chi = 1$ on the interval $[r_0, \infty)$.

Let us fix a positive real number $\bar{r}$ such that $\bar{r}$ is strictly smaller than the injectivity radius at $\bar{x}$. We define a smooth function $\rho : B(\bar{x}, \bar{r}) \to \mathbb{R}$ as follows. Given a point $x \in B(\bar{x}, \bar{r})$, there exists a unique vector $w \in T_{\bar{x}}M$ such that $|w| < \bar{r}$ and $x = \exp_{\bar{x}}(w)$. We denote by W the unique parallel vector field along γ satisfying $W(l) = w$. We then define $\sqrt{\rho(x)}$ to be the length of the curve

$$s \mapsto \exp_{\gamma(s)}(\chi(s)\, W(s)), \quad s \in [0, l].$$

Clearly,

$$\sqrt{\rho(x)} \geq d\big(\exp_{\gamma(0)}(\chi(0)\, W(0)), \exp_{\gamma(l)}(\chi(l)\, W(l))\big) = d(p, \exp_{\bar{x}}(w)) = d(p, x).$$

Moreover, in the special case when $x = \bar{x}$ and $w = 0$, we obtain

$$\sqrt{\rho(\bar{x})} = l = d(p, \bar{x}).$$

The formula for the first variation of arclength implies that $\nabla\sqrt{\rho} = \gamma'(l)$ at the point $\bar{x}$. In particular, $|\nabla\sqrt{\rho}|^2 = 1$ at the point $\bar{x}$. Consequently, $|\nabla\rho|^2 = 4\rho$ at the point $\bar{x}$.

Using the formula for the second variation of arclength, we obtain

$$(D^2\sqrt{\rho})_{\bar{x}}(w, w) = \int_0^l \chi'(s)^2 \langle W(s), W(s) \rangle \, ds - \int_0^l \chi'(s)^2 \langle \gamma'(s), W(s) \rangle^2 \, ds$$
$$- \int_0^l \chi(s)^2 R(\gamma'(s), W(s), \gamma'(s), W(s)) \, ds$$

for every vector $w \in T_{\bar{x}}M$, where W denotes the unique parallel vector field along γ satisfying $W(l) = w$. Taking the trace over w gives

$$\Delta\sqrt{\rho}(\bar{x}) = (n - 1) \int_0^l \chi'(s)^2 \, ds - \int_0^l \chi(s)^2 \, \mathrm{Ric}(\gamma'(s), \gamma'(s)) \, ds.$$

Since $\chi = 1$ on the interval $[r_0, l]$, we obtain

$$\Delta\sqrt{\rho}(\bar{x}) \leq -\int_0^l \mathrm{Ric}(\gamma'(s), \gamma'(s)) \, ds + C,$$

where C denotes a uniform constant that does not depend on $\bar{x}$.

On the other hand, using the identity $D^2 f = \mathrm{Ric}$, we obtain

$$\frac{d}{ds} \langle \nabla f(\gamma(s)), \gamma'(s) \rangle = \mathrm{Ric}(\gamma'(s), \gamma'(s)).$$

Integrating this identity over $s \in [0, l]$ gives

$$\langle \nabla f(\gamma(l)), \gamma'(l) \rangle \leq \int_0^l \mathrm{Ric}(\gamma'(s), \gamma'(s)) \, ds + C,$$

where C is a uniform constant that does not depend on $\bar{x}$. Since $\gamma(l) = \bar{x}$ and $\gamma'(l) = \nabla\sqrt{\rho}(\bar{x})$, we obtain

$$\langle \nabla f(\bar{x}), \nabla\sqrt{\rho}(\bar{x}) \rangle \leq \int_0^l \mathrm{Ric}(\gamma'(s), \gamma'(s)) \, ds + C,$$

where C is a uniform constant that does not depend on $\bar{x}$. Putting these facts together, we conclude that

$$\Delta\sqrt{\rho}(\bar{x}) + \langle \nabla f(\bar{x}), \nabla\sqrt{\rho}(\bar{x}) \rangle \leq C,$$

where C is a uniform constant that does not depend on $\bar{x}$. This finally implies

$$\Delta\rho(\bar{x}) + \langle \nabla f(\bar{x}), \nabla\rho(\bar{x})\rangle \leq 2 + C\sqrt{\rho},$$

where C is a uniform constant that does not depend on $\bar{x}$. This completes the proof of Lemma 8.1.

Proposition 8.2 (B.L. Chen [14]) *The manifold (M, g) has nonnegative scalar curvature.*

Proof As above, we fix an arbitrary point $p \in M$. Let N_0 and N_1 denote the constants in Lemma 8.1. Let us fix a radius $r > 0$. We define a continuous function $u : B(p, r) \to \mathbb{R}$ by

$$u(x) := R(x) + 2n(12 + N_0 + N_1 r)r^2(r^2 - d(p, x)^2)^{-2}$$

for $x \in B(p, r)$. We claim that $u(x) \geq 0$ for all $x \in B(p, r)$. To prove this, we argue by contradiction. Let $\bar{x}$ be a point in $B(p, r)$ where the function u attains its minimum, and suppose that $u(\bar{x}) < 0$. The evolution equation for the scalar curvature implies

$$\Delta R + \langle X, \nabla R\rangle = -2 \,|\,\mathrm{Ric}\,|^2.$$

By Lemma 8.1, we can find a smooth function ρ which is defined in an open neighborhood of $\bar{x}$ and satisfies the following conditions:

- $\rho(x) \geq d(p, x)^2$ in an open neighborhood of the point $\bar{x}$.
- $\rho(\bar{x}) = d(p, \bar{x})^2$.
- $|\nabla\rho|^2 = 4\rho$ at the point $\bar{x}$.
- $\Delta\rho + \langle X, \nabla\rho\rangle \leq N_0 + N_1\sqrt{\rho}$ at the point $\bar{x}$.

Then
$$R(x) + 2n(12 + N_0 + N_1 r)r^2(r^2 - \rho(x))^{-2} \geq u(x) \geq u(\bar{x})$$

in an open neighborhood of $\bar{x}$, with equality at the point $\bar{x}$. Consequently, the function $R + 2n(12 + N_0 + N_1 r)r^2(r^2 - \rho)^{-2}$ attains a local minimum at the point $\bar{x}$. Thus, we conclude that

$$
\begin{aligned}
0 &\leq \Delta R + \langle X, \nabla R\rangle \\
&\quad + 2n(12 + N_0 + N_1 r)r^2 \left[\Delta((r^2 - \rho)^{-2}) + \langle X, \nabla((r^2 - \rho)^{-2})\rangle\right] \\
&= \Delta R + \langle X, \nabla R\rangle \\
&\quad + 4n(12 + N_0 + N_1 r)r^2(r^2 - \rho)^{-3} \left(\Delta\rho + \langle X, \nabla\rho\rangle\right) \\
&\quad + 12n(12 + N_0 + N_1 r)r^2(r^2 - \rho)^{-4} |\nabla\rho|^2 \\
&\leq -2 \,|\,\mathrm{Ric}\,|^2 + 4n(12 + N_0 + N_1 r)(N_0 + N_1\sqrt{\rho})r^2(r^2 - \rho)^{-3} \\
&\quad + 48n(12 + N_0 + N_1 r)\rho r^2(r^2 - \rho)^{-4}
\end{aligned}
$$

at the point $\bar{x}$. Since $u \leq 0$ at the point $\bar{x}$, we know that

$$-R \geq 2n(12 + N_0 + N_1 r) r^2 (r^2 - \rho)^{-2}$$

at the point $\bar{x}$. This implies

$$n \,|\,\text{Ric}|^2 \geq R^2 \geq 4n^2 (12 + N_0 + N_1 r)^2 r^4 (r^2 - \rho)^{-4}$$

at the point $\bar{x}$. Putting these facts together, we obtain

$$
\begin{aligned}
0 \leq\ & -2\,|\,\text{Ric}|^2 + 4n(12 + N_0 + N_1 r)(N_0 + N_1 \sqrt{\rho}) r^2 (r^2 - \rho)^{-3} \\
& + 48n(12 + N_0 + N_1 r)\rho r^2 (r^2 - \rho)^{-4} \\
\leq\ & -8n(12 + N_0 + N_1 r)^2 r^4 (r^2 - \rho)^{-4} \\
& + 4n(12 + N_0 + N_1 r)(N_0 + N_1 r) r^4 (r^2 - \rho)^{-4} \\
& + 48n(12 + N_0 + N_1 r) r^4 (r^2 - \rho)^{-4} \\
=\ & -4n(12 + N_0 + N_1 r)^2 r^4 (r^2 - \rho)^{-4}
\end{aligned}
$$

at the point $\bar{x}$. This is a contradiction.

Thus, we conclude that

$$R(x) + 2n(12 + N_0 + N_1 r) r^2 (r^2 - d(p, x)^2)^{-2} \geq 0$$

for all $x \in B(p, r)$. Sending $r \to \infty$ gives $R(x) \geq 0$ for each point $x \in M$. This completes the proof of Proposition 8.2.

Corollary 8.3 *There exists a large constant C such that $|\nabla f| \leq C$ at each point on M.*

Proof Since (M, g, f) is a a steady gradient soliton, the sum $R + |\nabla f|^2$ is constant. Since $R \geq 0$ by Proposition 8.2, we conclude that $|\nabla f|$ is uniformly bounded from above. This completes the proof of Corollary 8.3.

In the remainder of this section, we assume that M is three-dimensional. As in Sect. 1, we denote by $\lambda_1 \leq \lambda_2 \leq \lambda_3$ the eigenvalues of the tensor $R\, g_{ij} - 2\,\text{Ric}_{ij}$. Then $R = \lambda_1 + \lambda_2 + \lambda_3$. Since $R \geq 0$, it follows that $\lambda_3 \geq 0$ at each point on M.

Proposition 8.4 (B.L. Chen [14]) *Assume that $n = 3$. Then $k\lambda_1 + 2R \geq 0$ for every nonnegative integer k.*

Proof The proof is by induction on k. For $k = 0$, the assertion follows from Proposition 8.2.

We now turn to the inductive step. Suppose that $k \geq 1$ and $(k - 1)\lambda_1 + 2R \geq 0$. As above, we fix an arbitrary point $p \in M$. Let N_0 and N_1 denote the constants in Lemma 8.1. Let us fix a radius $r > 0$. We define a continuous function $v : B(p, r) \to \mathbb{R}$ by

$$v(x) := k\lambda_1(x) + 2R(x) + 4k(12 + N_0 + N_1 r)r^2(r^2 - d(p, x)^2)^{-2}$$

for $x \in B(p, r)$. We claim that $v(x) \geq 0$ for all $x \in B(p, r)$. To prove this, we argue by contradiction. Let $\bar{x}$ be a point in $B(p, r)$ where the function v attains its minimum, and suppose that $v(\bar{x}) < 0$. The evolution equation for the scalar curvature implies

$$\Delta R + \langle X, \nabla R \rangle = -2\,|\operatorname{Ric}|^2.$$

The evolution equation for the Ricci tensor gives

$$\Delta\lambda_1 + \langle X, \nabla\lambda_1 \rangle \leq -(\lambda_1^2 + \lambda_2\lambda_3),$$

where the inequality is understood in the barrier sense. More precisely, we can find a smooth function ψ which is defined in an open neighborhood of $\bar{x}$ and satisfies the following conditions:

- $\psi(x) \geq \lambda_1(x)$ in an open neighborhood of the point $\bar{x}$.
- $\psi(\bar{x}) = \lambda_1(\bar{x})$.
- $\Delta\psi + \langle X, \nabla\psi \rangle \leq -(\lambda_1^2 + \lambda_2\lambda_3)$ at the point $\bar{x}$.

For example, we may define $\psi := \operatorname{Ric}(\xi, \xi)$, where ξ is a smooth unit vector field satisfying $\operatorname{Ric}(\xi, \xi) = \lambda_1$ at $\bar{x}$; $D\xi = 0$ at $\bar{x}$; and $\Delta\xi = 0$ at $\bar{x}$.

By Lemma 8.1, we can find a smooth function ρ which is defined in an open neighborhood of $\bar{x}$ and satisfies the following conditions:

- $\rho(x) \geq d(p, x)^2$ in an open neighborhood of the point $\bar{x}$.
- $\rho(\bar{x}) = d(p, \bar{x})^2$.
- $|\nabla\rho|^2 = 4\rho$ at the point $\bar{x}$.
- $\Delta\rho + \langle X, \nabla\rho \rangle \leq N_0 + N_1\sqrt{\rho}$ at the point $\bar{x}$.

Then

$$k\psi(x) + 2R(x) + 4k(12 + N_0 + N_1 r)r^2(r^2 - \rho(x))^{-2} \geq v(x) \geq v(\bar{x})$$

in an open neighborhood of $\bar{x}$, with equality at the point $\bar{x}$. Consequently, the function $k\psi + 2R + 4k(12 + N_0 + N_1 r)r^2(r^2 - \rho)^{-2}$ attains a local minimum at the point $\bar{x}$. Thus, we conclude that

$$0 \le k \left(\Delta\psi + \langle X, \nabla\psi\rangle\right) + 2\left(\Delta R + \langle X, \nabla R\rangle\right)$$
$$+ 4k(12 + N_0 + N_1 r) r^2 \left[\Delta((r^2 - \rho)^{-2}) + \langle X, \nabla((r^2 - \rho)^{-2})\rangle\right]$$
$$= k\left(\Delta\psi + \langle X, \nabla\psi\rangle\right) + 2\left(\Delta R + \langle X, \nabla R\rangle\right)$$
$$+ 8k(12 + N_0 + N_1 r) r^2 (r^2 - \rho)^{-3} \left(\Delta\rho + \langle X, \nabla\rho\rangle\right)$$
$$+ 24k(12 + N_0 + N_1 r) r^2 (r^2 - \rho)^{-4} |\nabla\rho|^2$$
$$\le -k\left(\lambda_1^2 + \lambda_2\lambda_3\right) - 4\,|\operatorname{Ric}|^2$$
$$+ 8k(12 + N_0 + N_1 r)(N_0 + N_1\sqrt{\rho}) r^2 (r^2 - \rho)^{-3}$$
$$+ 96k(12 + N_0 + N_1 r)\rho r^2 (r^2 - \rho)^{-4}$$

at the point $\bar{x}$. Since $v \le 0$ at the point $\bar{x}$, we know that

$$-(k\lambda_1 + 2R) \ge 4k(12 + N_0 + N_1 r) r^2 (r^2 - \rho)^{-2}$$

at the point $\bar{x}$. Note that $R \ge 0$, $(k-1)\lambda_1 + 2R \ge 0$, $k\lambda_1 + 2R \le 0$, $\lambda_1 \le 0$, and $\lambda_3 \ge 0$ at the point $\bar{x}$. This implies

$$k^2\left(\lambda_1^2 + \lambda_2\lambda_3\right) + 4k\,|\operatorname{Ric}|^2$$
$$= k^2\left(\lambda_1^2 + \lambda_2\lambda_3\right) + 2k\left(\lambda_1^2 + \lambda_2^2 + \lambda_3^2 + \lambda_1\lambda_2 + \lambda_2\lambda_3 + \lambda_3\lambda_1\right)$$
$$= (k\lambda_1 + 2R)^2 - 2R(k\lambda_1 + 2R) + k((k-1)\lambda_1 + 2R)\lambda_3$$
$$+ k^2(\lambda_2 - \lambda_1)\lambda_3 - k\lambda_1\lambda_3 + 2k\lambda_2^2$$
$$\ge (k\lambda_1 + 2R)^2$$
$$\ge 16k^2(12 + N_0 + N_1 r)^2 r^4 (r^2 - \rho)^{-4}$$

at the point $\bar{x}$. Putting these facts together, we obtain

$$0 \le -k\left(\lambda_1^2 + \lambda_2\lambda_3\right) - 4\,|\operatorname{Ric}|^2$$
$$+ 8k(12 + N_0 + N_1 r)(N_0 + N_1\sqrt{\rho}) r^2 (r^2 - \rho)^{-3}$$
$$+ 96k(12 + N_0 + N_1 r)\rho r^2 (r^2 - \rho)^{-4}$$
$$\le -16k(12 + N_0 + N_1 r)^2 r^4 (r^2 - \rho)^{-4}$$
$$+ 8k(12 + N_0 + N_1 r)(N_0 + N_1 r) r^4 (r^2 - \rho)^{-4}$$
$$+ 96k(12 + N_0 + N_1 r) r^4 (r^2 - \rho)^{-4}$$
$$= -8k(12 + N_0 + N_1 r)^2 r^4 (r^2 - \rho)^{-4}$$

at the point $\bar{x}$. This is a contradiction.

Thus, we conclude that

$$k\lambda_1(x) + 2R(x) + 4k(12 + N_0 + N_1 r) r^2 (r^2 - d(p, x)^2)^{-2} \ge 0$$

for all $x \in B(p, r)$. Sending $r \to \infty$ gives $k\lambda_1(x) + 2R(x) \geq 0$ for each point $x \in M$. This completes the proof of Proposition 8.4.

Corollary 8.5 (B.L. Chen [14]) *Assume that $n = 3$. Then (M, g) has nonnegative sectional curvature.*

Proof Sending $k \to \infty$ in Proposition 8.4 gives $\lambda_1 \geq 0$.

Corollary 8.6 *Assume that $n = 3$. Then (M, g) has bounded curvature.*

Proof Since (M, g, f) is a a steady gradient soliton, the sum $R + |\nabla f|^2$ is constant. Consequently, the scalar curvature is uniformly bounded from above. Hence, the assertion follows from Corollary 8.5.

From now on, we will assume that (M, g) is κ-noncollapsed. Moreover, we will assume that (M, g, f) is normalized so that $R + |\nabla f|^2 = 1$ at each point on M. Let Φ_t denote the one-parameter group of diffeomorphisms generated by the vector field $-X$. It follows from Corollary 8.3 that Φ_t is defined for all $t \in (-\infty, \infty)$. In view of Corollaries 8.5 and 8.6, the metrics $\Phi_t^*(g)$, $t \in (-\infty, 0]$, form an ancient κ-solution to the Ricci flow.

Proposition 8.7 *Assume that $n = 3$ and (M, g) is κ-noncollapsed. Then (M, g) has positive sectional curvature.*

Proof By Corollary 8.5, (M, g) has nonnegative sectional curvature. We claim that (M, g) has strictly positive sectional curvature. Suppose this is false. By the strict maximum principle, the universal cover of (M, g) splits off a line. Using Perelman's classification of ancient κ-solutions in dimension 2 (see Theorem 3.1), we conclude that the universal cover of (M, g) is isometric to a cylinder $S^2 \times \mathbb{R}$, up to scaling. In particular, (M, g) has constant scalar curvature. This contradicts the fact that $\Delta R + \langle X, \nabla R \rangle = -2 |\operatorname{Ric}|^2$ at each point on M. This completes the proof of Proposition 8.7.

Proposition 8.8 *Assume that $n = 3$ and (M, g) is κ-noncollapsed. Let p_j be a sequence of points going to infinity, and let $r_j^{-2} := R(p_j)$. Let us dilate the manifold (M, g) around the point p_j by the factor r_j^{-1}. Then, after passing to a subsequence if necessary, the rescaled manifolds converge in the Cheeger-Gromov sense to a cylinder of radius $\sqrt{2}$.*

Proof Since (M, g) has positive sectional curvature, the assertion follows from Corollary 4.4.

Corollary 8.9 *Assume that $n = 3$ and (M, g) is κ-noncollapsed. Then $R \to 0$ at infinity.*

Proof Proposition 8.8 implies that $R^{-2} \Delta R \to 0$ and $R^{-\frac{3}{2}} |\nabla R| \to 0$ at infinity. Since R is bounded from above, it follows that $\Delta R \to 0$ and $|\nabla R| \to 0$ at infinity. Since $|X|$ is bounded, we conclude that $\Delta R + \langle X, \nabla R \rangle \to 0$ at infinity. On the other hand, the evolution equation for the scalar curvature gives $\Delta R + \langle X, \nabla R \rangle = -2 |\operatorname{Ric}|^2$ at each point in M. Putting these facts together, we conclude that $|\operatorname{Ric}|^2 \to 0$ at infinity. This completes the proof of Corollary 8.9.

Corollary 8.10 *Assume that $n = 3$ and (M, g) is κ-noncollapsed. Let p_j be a sequence of points going to infinity, and let $r_j^{-2} := R(p_j)$. Let us dilate the manifold (M, g) around the point p_j by the factor r_j^{-1}. Then, after passing to a subsequence if necessary, the rescaled manifolds converge in the Cheeger-Gromov sense to a cylinder of radius $\sqrt{2}$, and the rescaled vector fields $r_j X$ converge in C_{loc}^∞ to the axial vector field on the cylinder.*

Proof The vector field X satisfies the pointwise estimates $|X| \leq 1$ and $|DX| = |\operatorname{Ric}| \leq C R$. Moreover, Perelman's pointwise derivative estimate (see Theorem 4.1) implies $|D^{m+1}X| = |D^m \operatorname{Ric}| \leq C R^{\frac{m+2}{2}}$ for every positive integer m. Consequently, the rescaled vector fields $r_j X$ converge in C_{loc}^∞ to a limit vector field on the cylinder. Since $|X|^2 = 1 - R \to 1$ at infinity, the limiting vector field on the cylinder has unit length at each point. Since $|DX| \leq C R$, the limiting vector field on the cylinder is parallel. This completes the proof of Corollary 8.10.

Proposition 8.11 *Assume that $n = 3$ and (M, g) is κ-noncollapsed. Then the function f has a unique critical point p_*, and f attains its global minimum at the point p_*.*

Proof In view of Corollary 8.9, there exists a point p_* where the scalar curvature is maximal. In particular, $\nabla R = 0$ at the point p_*. Since $R + |X|^2$ is constant, we know that $\nabla R + 2 \operatorname{Ric}(X) = 0$ at each point on M. Consequently, $\operatorname{Ric}(X) = 0$ at the point p_*. Since (M, g) has positive Ricci curvature, it follows that $X = 0$ at the point p_*. In other words, p_* is a critical point of f. Since (M, g) has positive Ricci curvature, the function f is strictly convex. Thus, p_* is the only critical point of f, and f attains its global minimum at the point p_*. This completes the proof of Proposition 8.11.

Corollary 8.12 *Assume that $n = 3$ and (M, g) is κ-noncollapsed. Then there exists a positive constant C such that*

$$\frac{1}{C} d(p_*, x) \leq f(x) \leq C d(p_*, x)$$

outside some compact set.

Proof The upper bound for f follows from Corollary 8.3. The lower bound follows from Proposition 8.11 together with the strict convexity of f.

Proposition 8.13 (H. Guo [20]) *Assume that $n = 3$ and (M, g) is κ-noncollapsed. Then $f R \to 1$ at infinity.*

Proof Using the evolution equation for the scalar curvature, we obtain

$$\Delta R + \langle X, \nabla R \rangle = -2 |\operatorname{Ric}|^2$$

at each point in M. This implies

$$\langle X, \nabla(R^{-1} - f)\rangle = -R^{-2} \langle X, \nabla R\rangle - |\nabla f|^2$$
$$= R^{-2} \Delta R + 2R^{-2} | \operatorname{Ric}|^2 - |\nabla f|^2$$

at each point in M. Using Proposition 8.8, we obtain $R^{-2} \Delta R \to 0$ and $R^{-2} | \operatorname{Ric}|^2 \to \frac{1}{2}$ at infinity. Moreover, Corollary 8.9 implies $|\nabla f|^2 = 1 - R \to 1$ at infinity. Putting these facts together, we conclude that

$$\langle X, \nabla(R^{-1} - f)\rangle \to 0$$

at infinity. Hence, if $\varepsilon > 0$ is given, then

$$\langle X, \nabla(R^{-1} - (1+\varepsilon)f)\rangle \le 0$$

and

$$\langle X, \nabla(R^{-1} - (1-\varepsilon)f)\rangle \ge 0$$

outside a compact set. Integrating these inequalities along the integral curves of X, we obtain

$$\sup_M (R^{-1} - (1+\varepsilon)f) < \infty$$

and

$$\inf_M (R^{-1} - (1-\varepsilon)f) > -\infty.$$

Since $\varepsilon > 0$ is arbitrary, we conclude that $f R \to 1$ at infinity. This completes the proof of Proposition 8.13.

In particular, if s is sufficiently large and $f(x) = s$, then x lies at center of a neck, and the radius of the neck is $(1 + o(1))\sqrt{2s}$.

9 Rotational Symmetry of Noncollapsed Steady Gradient Ricci Solitons in Dimension 3 – The Proof of Theorem 5.1

Throughout this section, we assume that (M, g, f) is a non-flat steady gradient Ricci soliton in dimension 3 which is κ-noncollapsed. Moreover, we assume that (M, g, f) is normalized so that $R + |\nabla f|^2 = 1$ at each point on M. It follows from Proposition 8.11 that f is bounded from below. After adding a constant to f if necessary, we may assume that f is positive at each point on M. For abbreviation, we put $X := \nabla f$. Let Φ_t denote the one-parameter group of diffeomorphisms generated by the vector field $-X$. By Corollary 8.3, Φ_t is defined for all $t \in (-\infty, \infty)$. Moreover, the metrics $\Phi_t^*(g)$, $t \in (-\infty, 0]$, form an ancient κ-solution to the Ricci flow.

Let us fix a large real number L and a small positive real number ε_1 so that the conclusion of the Neck Improvement Theorem holds. In view of Proposition 8.13, we can find a large constant Λ with the following properties:

- $f(x) R(x, 0)^{\frac{1}{2}} \geq 10^6 L$ for each point $x \in M$ with $f(x) \geq \frac{\Lambda}{2}$.
- If $f(\bar{x}) \geq \frac{\Lambda}{2}$, then $(\bar{x}, 0)$ lies at the center of an evolving ε_1^2-neck.

By a repeated application of the Neck Improvement Theorem, we obtain the following result.

Proposition 9.1 *Suppose that j is a nonnegative integer and x is a point in M with $f(x) \geq 2^{\frac{j}{400}} \Lambda$. Then the point $(x, 0)$ is $2^{-j} \varepsilon_1$-symmetric.*

Proof The proof is by induction on j. For $j = 0$, the assertion is true by our choice of Λ. Suppose next that $j \geq 1$, and the assertion is true for $j - 1$. Let us fix a point $\bar{x} \in M$ satisfying $f(\bar{x}) \geq 2^{\frac{j}{400}} \Lambda$, and let $r^{-2} := R(\bar{x}, 0)$. By our choice of Λ, $f(\bar{x}) r^{-1} = f(\bar{x}) R(\bar{x}, 0)^{\frac{1}{2}} \geq 10^6 L$. Since $|\nabla f| \leq 1$ at each point on M, we obtain

$$f(x) \geq f(\bar{x}) - Lr \geq (1 - 10^{-6}) f(\bar{x}) \geq (1 - 10^{-6}) 2^{\frac{j}{400}} \Lambda \geq 2^{\frac{j-1}{400}} \Lambda$$

for each point $x \in B_g(\bar{x}, Lr)$. This implies

$$f(\Phi_t(x)) \geq f(x) \geq 2^{\frac{j-1}{400}} \Lambda$$

for each point $x \in B_g(\bar{x}, Lr)$ and each $t \leq 0$. We now apply the induction hypothesis. Hence, if $x \in B_g(\bar{x}, Lr)$ and $t \leq 0$, then the point $(\Phi_t(x), 0)$ is $2^{-j+1} \varepsilon_1$-symmetric. Since we are working on a self-similar solution, the point (x, t) plays the same role as the point $(\Phi_t(x), 0)$. Consequently, if $x \in B_g(\bar{x}, Lr)$ and $t \leq 0$, then the point (x, t) is $2^{-j+1} \varepsilon_1$-symmetric. Using the Neck Improvement Theorem, we conclude that the point $(\bar{x}, 0)$ is $2^{-j} \varepsilon_1$-symmetric. This completes the proof of Proposition 9.1.

Corollary 9.2 *If m is sufficiently large, then we can find smooth vector fields $U^{(1,m)}, U^{(2,m)}, U^{(3,m)}$ on the domain $\{m - 80\sqrt{m} \leq f \leq m + 80\sqrt{m}\}$ with the following properties:*

- $\sum_{l=0}^{2} \sum_{a=1}^{3} m^l |D^l(\mathscr{L}_{U^{(a,m)}}(g))|^2 \leq C m^{-800}$.
- *If $\Sigma \subset \{m - 80\sqrt{m} \leq f \leq m + 80\sqrt{m}\}$ is a leaf of the CMC foliation, then $\sup_{\Sigma} \sum_{a=1}^{3} m^{-1} |\langle U^{(a)}, \nu \rangle|^2 \leq C m^{-800}$, where ν denotes the unit normal vector to Σ.*
- *If $\Sigma \subset \{m - 80\sqrt{m} \leq f \leq m + 80\sqrt{m}\}$ is a leaf of the CMC foliation, then*

$$\sum_{a,b=1}^{3} \left| \delta_{ab} - area(\Sigma)^{-2} \int_{\Sigma} \langle U^{(a,m)}, U^{(b,m)} \rangle \, d\mu \right|^2 \leq C m^{-800}.$$

As explained in Sect. 7 of [8], we can glue approximate Killing vector fields on overlapping necks. This allows us to draw the following conclusion:

Corollary 9.3 *We can find smooth vector fields $U^{(1)}$, $U^{(2)}$, $U^{(3)}$ such that*

$$|\mathscr{L}_{U^{(a)}}(g)| \le C\,(f + 100)^{-100},$$

$$|D(\mathscr{L}_{U^{(a)}}(g))| \le C\,(f + 100)^{-100},$$

$$|D^2(\mathscr{L}_{U^{(a)}}(g))| \le C\,(f + 100)^{-100}.$$

Finally, given any positive real number ε, we have

$$\sum_{a,b=1}^{3} \left| \delta_{ab} - area(\Sigma)^{-2} \int_{\Sigma} \langle U^{(a)}, U^{(b)} \rangle \, d\mu \right|^2 \le \varepsilon^2$$

whenever Σ is a leaf of the CMC foliation which is sufficiently far out near infinity (depending on ε).

Lemma 9.4 *We have $|\langle [U^{(a)}, X], X \rangle| \le C\,(f + 100)^{-40}$.*

Proof The vector field X satisfies $|X|^2 = 1 - R$. Let us take the Lie derivative along $U^{(a)}$ on both sides. This gives

$$(\mathscr{L}_{U^{(a)}}(g))(X, X) + 2\,\langle \mathscr{L}_{U^{(a)}}(X), X \rangle = -\mathscr{L}_{U^{(a)}}(R).$$

Using the formula for the linearization of the scalar curvature (see [4], Theorem 1.174 (e)), we obtain

$$|\mathscr{L}_{U^{(a)}}(R)| \le C\,|D^2(\mathscr{L}_{U^{(a)}}(g))| + C\,|\,\text{Ric}|\,|\mathscr{L}_{U^{(a)}}(g)| \le C\,(f + 100)^{-40}.$$

Putting these facts together, we conclude that

$$|\langle \mathscr{L}_{U^{(a)}}(X), X \rangle| \le C\,(f + 100)^{-40}.$$

This completes the proof of Lemma 9.4.

Lemma 9.5 *We have $|D([U^{(a)}, X])| \le C\,(f + 100)^{-40}$.*

Proof The vector field X satisfies $g_{jk}\,D_i X^j = \text{Ric}_{ik}$. Let us take the Lie derivative along $U^{(a)}$ on both sides. Using the formula for the linearization of the Levi-Civita connection (see [4], Theorem 1.174 (a)), we obtain

$$(\mathscr{L}_{U^{(a)}}(g))_{jk}\,D_i X^j + g_{jk}\,D_i(\mathscr{L}_{U^{(a)}}(X))^j$$
$$+ \frac{1}{2}\,D_i(\mathscr{L}_{U^{(a)}}(g))_{jk}\,X^j + \frac{1}{2}\,D_j(\mathscr{L}_{U^{(a)}}(g))_{ik}\,X^j - \frac{1}{2}\,D_k(\mathscr{L}_{U^{(a)}}(g))_{ij}\,X^j$$
$$= (\mathscr{L}_{U^{(a)}}(\text{Ric}))_{ik}.$$

The formula for the linearization of the Ricci tensor (see [4], Theorem 1.174 (d)) gives

$$|\mathcal{L}_{U^{(a)}}(\,\mathrm{Ric})| \leq C\,|D^2(\mathcal{L}_{U^{(a)}}(g))| + C\,|\,\mathrm{Rm}|\,|\mathcal{L}_{U^{(a)}}(g)| \leq C\,(f+100)^{-40}.$$

Putting these facts together, we conclude that

$$|D(\mathcal{L}_{U^{(a)}}(X))| \leq C\,(f+100)^{-40}.$$

This completes the proof of Lemma 9.5.

Lemma 9.6 *We have* $\|[U^{(a)}, X]\| \leq (f+100)^{-20}$ *outside a compact set.*

Proof Suppose that the assertion is false. Then there exists an index $a \in \{1, 2, 3\}$ and a sequence of points p_j going to infinity such that

$$\|[U^{(a)}, X]\| \geq (f+100)^{-20}$$

at the point p_j. Let us define $s_j := f(p_j)$, and let A_j denote the norm of the vector field $[U^{(a)}, X]$ at the point p_j. By assumption, $A_j \geq (s_j + 100)^{-20}$ for each j. Since $|\nabla f| \leq 1$ at each point on M, we know that $\frac{s_j}{2} \leq f \leq \frac{3s_j}{2}$ at each point in $B_g(p_j, \frac{s_j}{2})$. Using Lemmas 9.4 and 9.5, we obtain

$$\sup_{B_g(p_j, \frac{s_j}{2})} |\langle [U^{(a)}, X], X \rangle| \leq C\,(s_j + 100)^{-40} \leq C\,(s_j + 100)^{-20}\,A_j$$

and

$$\sup_{B_g(p_j, \frac{s_j}{2})} |D([U^{(a)}, X])| \leq C\,(s_j + 100)^{-40} \leq C\,(s_j + 100)^{-20}\,A_j.$$

In the next step, we integrate the bound for $D([U^{(a)}, X])$ along geodesics emanating from p_j. If j is sufficiently large, we obtain

$$\sup_{B_g(p_j, \frac{s_j}{2})} \|[U^{(a)}, X]\| \leq A_j + C\,(s_j + 100)^{-30} \leq 2A_j.$$

We now dilate the manifold (M, g) around the point p_j by the factor $s_j^{-\frac{1}{2}}$. By Corollary 8.10, the rescaled manifolds converge in the Cheeger-Gromov sense to a cylinder of radius $\sqrt{2}$, and the rescaled vector fields $s_j^{\frac{1}{2}} X$ converge in C^∞_{loc} to the axial vector field on the cylinder. Moreover, the vector fields $s_j^{\frac{1}{2}} A_j^{-1} [U^{(a)}, X]$ converge in $C^{\frac{1}{2}}_{\mathrm{loc}}$ to a non-trivial parallel vector field on the cylinder, and this limiting vector field is orthogonal to the axial vector field on the cylinder. This is a contradiction. This completes the proof of Lemma 9.6.

Lemma 9.7 *We have* $|\Delta U^{(a)} + D_X U^{(a)}| \leq C\,(f+100)^{-20}$.

Proof Using Proposition 6.2 and Corollary 9.3, we obtain

$$|\Delta U^{(a)} + \text{Ric}(U^{(a)})| \le C\,|D(\mathscr{L}_{U^{(a)}}(g))| \le C\,(f+100)^{-100}.$$

Moreover, Lemma 9.6 gives

$$|[U^{(a)}, X]| \le C\,(f+100)^{-20}.$$

Using the identity

$$\begin{aligned}
\Delta U^{(a)} + D_X U^{(a)} &= \Delta U^{(a)} + D_{U^{(a)}} X - [U^{(a)}, X] \\
&= \Delta U^{(a)} + \text{Ric}(U^{(a)}) - [U^{(a)}, X],
\end{aligned}$$

we conclude that
$$|\Delta U^{(a)} + D_X U^{(a)}| \le C\,(f+100)^{-20}.$$

This completes the proof of Lemma 9.7.

For abbreviation, we define smooth vector fields $Q^{(1)}, Q^{(2)}, Q^{(3)}$ by $Q^{(a)} := \Delta U^{(a)} + D_X U^{(a)}$.

Proposition 9.8 *We can find smooth vector fields $W^{(1)}, W^{(2)}, W^{(3)}$ such that $|W^{(a)}| \le C\,(f+100)^{-8}$ and $\Delta W^{(a)} + D_X W^{(a)} = Q^{(a)}$.*

Proof We consider a sequence of real numbers $s_j \to \infty$. For each j and each $a \in \{1, 2, 3\}$, we denote by $W^{(a,j)}$ the solution of the elliptic PDE

$$\Delta W^{(a,j)} + D_X W^{(a,j)} = Q^{(a)}$$

on the domain $\{f \le s_j\}$ with Dirichlet boundary condition $W^{(a,j)} = 0$ on the boundary $\{f = s_j\}$. This Dirichlet problem has a solution by the Fredholm alternative. Moreover, since $Q^{(a)}$ is smooth, it follows that $W^{(a,j)}$ is smooth.

By Lemma 9.7, $Q^{(a)}$ satisfies a pointwise estimate of the form

$$|Q^{(a)}| \le K\,(f+100)^{-20},$$

where K is a large constant that does not depend on j. Using Kato's inequality, we obtain

$$\Delta |W^{(a,j)}| + \langle X, \nabla|W^{(a,j)}|\rangle \ge -|Q^{(a)}| \ge -K\,(f+100)^{-20}$$

on the set $\{W^{(a,j)} \ne 0\}$. On the other hand, using the identity $\Delta f + \langle X, \nabla f\rangle = R + |\nabla f|^2 = 1$, we obtain

$$\Delta((f + 100)^{-8}) + \langle X, \nabla((f + 100)^{-8}) \rangle$$
$$= -8 (f + 100)^{-9} (\Delta f + \langle X, \nabla f \rangle) + 72 (f + 100)^{-10} |\nabla f|^2$$
$$\leq -8 (f + 100)^{-9} + 72 (f + 100)^{-10}$$
$$\leq -(f + 100)^{-9}.$$

Using the maximum principle, we conclude that

$$|W^{(a,j)}| \leq K (f + 100)^{-8}$$

on the set $\{f \leq s_j\}$.

We now send $j \to \infty$. After passing to a subsequence, the vector fields $W^{(a,j)}$ converge in C^∞_{loc} to a smooth vector field $W^{(a)}$. The limiting vector field $W^{(a)}$ satisfies

$$|W^{(a)}| \leq K (f + 100)^{-8}$$

and

$$\Delta W^{(a)} + D_X W^{(a)} = Q^{(a)}.$$

This completes the proof of Proposition 9.8.

Proposition 9.9 *We have* $|DW^{(a)}| \leq C (f + 100)^{-8}$.

Proof By Proposition 9.8, the vector field $W^{(a)}$ satisfies $\Delta W^{(a)} + D_X W^{(a)} = Q^{(a)}$. This equation can be rewritten as $\Delta W^{(a)} + \mathscr{L}_X(W^{(a)}) + \text{Ric}(W^{(a)}) = Q^{(a)}$. We next consider the vector fields $\Phi^*_t(W^{(a)})$ on the evolving background $(M, \Phi^*_t(g))$. These vector fields satisfy the parabolic PDE

$$\frac{\partial}{\partial t} \Phi^*_t(W^{(a)}) = \Delta_{\Phi^*_t(g)} \Phi^*_t(W^{(a)}) + \text{Ric}_{\Phi^*_t(g)}(\Phi^*_t(W^{(a)})) - \Phi^*_t(Q^{(a)}).$$

The assertion follows now from interior estimates for parabolic PDE (see e.g. [11], Proposition C.2). This completes the proof of Proposition 9.9.

We next define smooth vector fields $V^{(1)}, V^{(2)}, V^{(3)}$ by $V^{(a)} := U^{(a)} - W^{(a)}$.

Proposition 9.10 *The vector field* $V^{(a)}$ *satisfies* $\Delta V^{(a)} + D_X V^{(a)} = 0$.

Proof This follows immediately from Proposition 9.8.

Proposition 9.11 *The tensor* $\mathscr{L}_{V^{(a)}}(g)$ *vanishes identically.*

Proof Recall that

$$\Delta V^{(a)} + D_X V^{(a)} = 0.$$

By Corollary 6.4, the tensor $h^{(a)} := \mathscr{L}_{V^{(a)}}(g)$ satisfies

$$\Delta_L h^{(a)} + \mathscr{L}_X(h^{(a)}) = 0.$$

Hence, Corollary 6.6 implies

$$\Delta\left(\frac{|h^{(a)}|^2}{R^2}\right) + \left\langle X, \nabla\left(\frac{|h^{(a)}|^2}{R^2}\right)\right\rangle + \frac{2}{R}\left\langle \nabla R, \nabla\left(\frac{|h^{(a)}|^2}{R^2}\right)\right\rangle \geq 0.$$

Using the maximum principle, we obtain

$$\sup_{\{f \leq s\}} \frac{|h^{(a)}|}{R} \leq \sup_{\{f=s\}} \frac{|h^{(a)}|}{R}$$

for each s. On the other hand, Corollary 9.3 and Proposition 9.9 imply

$$|h^{(a)}| \leq |\mathcal{L}_{U^{(a)}}(g)| + C\,|DW^{(a)}| \leq C\,(f+100)^{-8}.$$

Using Proposition 8.13, we deduce that

$$\frac{|h^{(a)}|}{R} \leq C\,(f+100)^{-7}.$$

In particular,

$$\sup_{\{f=s\}} \frac{|h^{(a)}|}{R} \to 0$$

as $s \to \infty$. Putting these facts together, we conclude that $h^{(a)}$ vanishes identically. This completes the proof of Proposition 9.11.

Corollary 9.12 *We have* $[V^{(a)}, X] = 0$ *and* $\langle V^{(a)}, X\rangle = 0$.

Proof Note that

$$\Delta V^{(a)} + D_X V^{(a)} = 0$$

by Proposition 9.10. On the other hand, since $V^{(a)}$ is a Killing vector field, we obtain

$$\Delta V^{(a)} + \mathrm{Ric}(V^{(a)}) = 0$$

by Proposition 6.2. This implies $D_X V^{(a)} = \mathrm{Ric}(V^{(a)})$. On the other hand, $D_{V^{(a)}} X = \mathrm{Ric}(V^{(a)})$. Consequently, $[V^{(a)}, X] = 0$. This proves the first statement. We now turn to the proof of the second statement. Since $V^{(a)}$ is a Killing vector field, we obtain

$$\nabla(\mathcal{L}_{V^{(a)}}(f)) = \mathcal{L}_{V^{(a)}}(\nabla f) = \mathcal{L}_{V^{(a)}}(X) = 0.$$

Consequently, the function $\mathcal{L}_{V^{(a)}}(f) = \langle V^{(a)}, X\rangle$ is constant. On the other hand, by Proposition 8.11, the vector field X vanishes at some point $p_* \in M$. Thus, we conclude that the function $\langle V^{(a)}, X\rangle$ vanishes identically. This completes the proof of Corollary 9.12.

Corollary 9.12 implies that the vector fields $V^{(1)}$, $V^{(2)}$, $V^{(3)}$ are tangential to the level sets of f.

Proposition 9.13 *Given any positive real number ε, we have*

$$\sum_{a,b=1}^{3} \left| \delta_{ab} - area(\Sigma)^{-2} \int_{\Sigma} \langle V^{(a)}, V^{(b)} \rangle \, d\mu \right|^2 \leq \varepsilon^2$$

whenever Σ is a leaf of the CMC foliation which is sufficiently far out near infinity (depending on ε).

Proof This follows by combining Corollary 9.3 with Proposition 9.8.

Proposition 9.13 ensures that the vector fields $V^{(1)}$, $V^{(2)}$, $V^{(3)}$ are non-trivial near infinity. From this, it is easy to see that (M, g) is rotationally symmetric.

References

1. G. Anderson, B. Chow, A pinching estimate for solutions of the linearized Ricci flow system on 3-manifolds. Calc. Var. PDE **23**, 1–12 (2005)
2. S. Angenent, S. Brendle, P. Daskalopoulos, N. Šešum, Unique asymptotics of compact ancient solutions to three-dimensional Ricci flow. Comm. Pure Appl. Math. **75**, 1032–1073 (2022)
3. S. Angenent, P. Daskalopoulos, N. Šešum, Uniqueness of two-convex closed ancient solutions to the mean curvature flow. Ann. Math. **192**, 353–436 (2020)
4. A.L. Besse, *Einstein Manifolds* (Springer, Berlin, 1987)
5. S. Brendle, A general convergence result for the Ricci flow in higher dimensions. Duke Math. J. **145**, 585–601 (2008)
6. S. Brendle, *Ricci Flow and the Sphere Theorem*, Graduate Studies in Mathematics, vol. 111. (American Mathematical Society, 2010)
7. S. Brendle, Rotational symmetry of self-similar solutions to the Ricci flow. Invent. Math. **194**, 731–764 (2013)
8. S. Brendle, Ancient solutions to the Ricci flow in dimension 3. Acta Math. **225**, 1–102 (2020)
9. S. Brendle, K. Choi, Uniqueness of convex ancient solutions to mean curvature flow in $\mathbb{R}^3$. Invent. Math. **217**, 35–76 (2019)
10. S. Brendle, P. Daskalopoulos, N. Šešum, Uniqueness of compact ancient solutions to three-dimensional Ricci flow. Invent. Math. **226**, 579–651 (2021)
11. S. Brendle, K. Naff, Rotational symmetry of ancient solutions to the Ricci flow in higher dimensions. To appear in Geom. Topol
12. S. Brendle, R. Schoen, Manifolds with 1/4-pinched curvature are space forms. J. Amer. Math. Soc. **22**, 287–307 (2009)
13. R.L. Bryant, *Ricci flow solitons in dimension three with $SO(3)$-symmetries*. www.math.duke.edu/~bryant/3DRotSymRicciSolitons.pdf
14. B.L. Chen, Strong uniqueness of the Ricci flow. J. Diff. Geom. **82**, 363–382 (2009)
15. B. Chow, The Ricci flow on the 2-sphere. J. Diff. Geom. **33**, 325–334 (1991)
16. B. Chow, *On a formula of Daskalopoulos, Hamilton, and Šešum*. arxiv:1209.1784
17. P. Daskalopoulos, R. Hamilton, N. Šešum, Classification of ancient compact solutions to the Ricci flow on surfaces. J. Diff. Geom. **91**, 171–214 (2012)
18. D. DeTurck, Deforming metrics in the direction of their Ricci tensors. J. Diff. Geom. **18**, 157–162 (1983)

19. J. Eells, L. Lemaire, A report on harmonic maps. Bull. Lond. Math. Soc. **10**, 1–68 (1978)
20. H. Guo, Area growth rate of the level surface of the potential function on the 3-dimensional steady Ricci soliton. Proc. Amer. Math. Soc. **137**, 2093–2097 (2009)
21. R. Hamilton, Three-manifolds with positive Ricci curvature. J. Diff. Geom. **17**, 255–306 (1982)
22. R. Hamilton, *The Ricci flow on surfaces*, Contemporary Mathematics, vol. 71. (American Mathematical Society, Providence RI, 1988), pp. 237–262
23. R. Hamilton, *The formation of singularities in the Ricci flow,* Surveys in Differential Geometry, vol. II. (International Press, Somerville MA, 1995), pp. 7–136
24. R. Hamilton, Four-manifolds with positive isotropic curvature. Comm. Anal. Geom. **5**, 1–92 (1997)
25. T. Ivey, Ricci solitons on compact three-manifolds. Diff. Geom. Appl. **3**, 301–307 (1993)
26. J.R. King, Exact polynomial solutions to some nonlinear diffusion equations. Physica D **64**, 39–65 (1993)
27. G. Perelman, *The entropy formula for the Ricci flow and its geometric applications.* arXiv:0211159
28. G. Perelman, *Ricci flow with surgery on three-manifolds.* arxiv:0303109
29. P. Rosenau, Fast and super fast diffusion processes. Phys. Rev. Lett. **74**, 1056–1059 (1995)
30. P. Topping, *Lectures on the Ricci Flow*, London Mathematical Society Lecture Notes Series, vol. 325. (Cambridge University Press, Cambridge, 2006)

Spectral Geometry and Analysis of the Neumann-Poincaré Operator, a Review

Hyeonbae Kang

Abstract The Neumann-Poincaré operator is an integral operator defined on the boundary of a bounded domain. The history of research on it goes back to the era of the mathematicians whose names appear on the name of the operator. The spectral theory of the Neumann-Poincaré operator attracts much attention lately mainly due to its connection to plasmon resonance and cloaking by anomalous localized resonance. There are rapidly growing literature of research results on its spectral geometry and analysis, and the purpose of this paper is to review some of them. Topics of review in this paper include cloaking by anomalous localized resonance and analysis of surface localization of plasmon, negative eigenvalues and spectrum on tori, spectrum on polygonal domains, spectral structure of thin domains, and analysis of stress in terms of spectral theory. These topics are chosen so that they do not overlap those in another review paper on the same subject [22]. We also discuss some related problems to be considered for further development.

Keywords Neumann-Poincaré operator · Spectral geometry · Spectral analysis · Plasmon resonance · Cloaking by anomalous localized resonance · Negative eigenvalue · Torus · Thin domain · Stress

1 Introduction

This paper is a survey on some of recent development on the spectral theory of the Neumann-Poincaré (abbreviated by NP) operator. The NP operator is an integral operator defined on the boundary of a bounded domains which appears naturally when solving classical Dirichlet or Neumann boundary value problems using layer

---

This work is partially supported by National Research Foundation (of S. Korea) grants No. 2019R1A2B5B01069967.

H. Kang (✉)
Department of Mathematics and Institute of Applied Mathematics, Inha University,
Incheon 22212, South Korea
e-mail: hbkang@inha.ac.kr

potentials. The history of research on the NP operator goes back to the era of C. Neumann and Poincaré as the name of the operator suggests. It gave birth to the Fredholm theory of integral equations. If the domain on which the NP operator is defined has a corner, then it is a singular integral operator of which the theory has been one of central themes of research of the last century. For example, its L^2-boundedness was proved in the seminal paper [29]. Lately, the spectral theory of the NP operator attracts much attention in connection with exotic physical phenomena such as plasmon resonance and cloaking by anomalous localized resonance.

The author has written another survey paper on the spectral theory of the NP operator with his coauthors [22]. In this paper we omit introductory remarks and more historical accounts leaving them to that paper. The topics of review in this paper are chosen so that they don't overlap those in that paper. Topics in [22] include

- essential spectrum,
- decay estimates of NP eigenvalues in two dimensions,
- Weyl-type asymptotic formula for eigenvalues in three dimensions,
- the elastic NP operator,
- spectral analysis in a space with two norms.

Topics in this paper are

- NP operator and plasmon resonance,
- cloaking by anomalous localized resonance and analysis of surface localization of plasmon,
- concavity and negative eigenvalues including NP spectral structure on tori,
- spectrum on polygonal domains (with emphasis on pure point spectrum),
- spectral structure of thin domains,
- analysis of stress in terms of the spectral theory.

These topics are complementary to each other. Be aware that the NP operator in [22] is 2 times the NP operator of this paper.

The plan of this paper is as follows. Recent rapid growth of interest in the NP spectrum is mainly due to its connection to plasmon resonance and cloaking by anomalous localized resonance (abbreviated by CALR). In Sect. 2 we explain plasmon resonance in quasi-static limit in terms of a transmission problem. The NP operator is naturally introduced in the course of explanation. In Sect. 3 we discuss the spectral nature of CALR in terms of surface localization of plasmon and the decay rate of NP eigenvalues (eigenvalues of the NP operator). We review CALR on ellipses and a recent proof of non-occurrence of CALR on strictly convex three-dimensional domains, and peculiar spectral properties on tori in relation with CALR. In Sect. 4 we review recent results on negative NP eigenvalues and concavity in three dimensions. We include a discussion on possible advantage of having negative eigenvalues. In Sect. 5 we review results on the NP spectrum on planar domains with corners and discuss existence of eigenvalues in addition to continuous spectrum. In Sect. 6 we review the spectral structure of thin domains in two and three dimensions which is related to negative eigenvalues and polygonal domains. Quantitative analysis of the stress or field concentration in between two closely located inclusions has been an active area of research for last thirty years or so. In Sect. 7 we review some of important results on this subject and discuss a spectral nature of stress concentration, especially when one inclusion is an insulator and the other is a perfect conductor. In the course of review, proper references will be given in each corresponding section and some related open problems are discussed. This paper ends with a conclusion.

2   NP Operator and Plasmon Resonance

In this section we discuss plasmon resonance, which occurs on meta-materials of negative dielectric constants, as a motivation to study the NP operator and its spectrum. In the course of discussion, the single layer potential and the NP operator appear naturally.

Let Ω be a bounded domain in $\mathbb{R}^d$ ($d = 2, 3$). The boundary $\partial\Omega$ is allowed to have several connected components and its connected component is assumed to be Lipschitz continuous. Suppose that Ω is immersed in the free space $\mathbb{R}^d$ and the dielectric constant of Ω is $\epsilon_c = k + i\delta$ and that of the background is 1 after normalization ($k \neq 1$); (k, δ are real constants and δ is the lossy parameter tending to 0). The constant k can be negative. A material whose dielectric constant has the negative real part is called a meta-material. So, the distribution of the conductivity or the dielectric constant is given by

$$\epsilon = \epsilon_c \chi(\Omega) + \chi(\mathbb{R}^d \setminus \overline{\Omega}), \tag{1}$$

where χ denotes the indicator function of the corresponding set. We consider the following transmission problem: for a given harmonic function h in $\mathbb{R}^d$

$$\begin{cases} \nabla \cdot \epsilon \nabla u = 0 & \text{in } \mathbb{R}^d, \\ u(x) - h(x) = O(|x|^{1-d}) & \text{as } |x| \to \infty. \end{cases} \tag{2}$$

The solution to the problem, denoted by u_δ, satisfies the transmission conditions along $\partial\Omega$:

$$u_\delta|_- = u_\delta|_+, \quad \epsilon_c \partial_\nu u_\delta|_- = \partial_\nu u_\delta|_+, \tag{3}$$

which are continuity of the potential and the flux. Here and afterwards, subscripts $+$ and $-$ indicate the limits (to $\partial\Omega$) from outside and inside of Ω, respectively, and ∂_ν the outward normal derivative on $\partial\Omega$.

The solution to (2) can be represented using the single layer potential which is defined, for $\varphi \in H^{-1/2}(\partial\Omega)$ ($H^{-1/2}(\partial\Omega)$ is the usual Sobolev space on $\partial\Omega$), by

$$\mathcal{S}_{\partial\Omega}[\varphi](x) := \int_{\partial\Omega} \Gamma(x - y)\varphi(y) \, d\sigma(y), \quad x \in \mathbb{R}^d, \tag{4}$$

where $\Gamma(x)$ is the fundamental solution to the Laplacian, i.e.,

$$\Gamma(x) = \begin{cases} \dfrac{1}{2\pi} \ln|x| \, , & d = 2 \, , \\[2mm] -\dfrac{1}{4\pi}|x|^{-1} \, , & d = 3 \, . \end{cases} \tag{5}$$

It turns out (see, for example, [10, 50]) that the solution u_δ to (2) can be represented as

$$u_\delta(x) = h(x) + \mathcal{S}_{\partial\Omega}[\varphi_\delta](x), \quad x \in \mathbb{R}^d, \tag{6}$$

for some $\varphi_\delta \in H_0^{-1/2}(\partial\Omega)$ (the subscript 0 indicates that its elements are of zero mean value). The potential function φ_δ is determined by the transmission condition (3). The first condition (continuity of the potential) is automatically fulfilled since $\mathcal{S}_{\partial\Omega}[\varphi_\delta]$ is continuous across $\partial\Omega$. The second condition (continuity of the flux) takes the following form:

$$\epsilon_c \partial_\nu \mathcal{S}_{\partial\Omega}[\varphi_\delta]|_- - \partial_\nu \mathcal{S}_{\partial\Omega}[\varphi_\delta]|_+ = (1 - \epsilon_c)\partial_\nu h. \tag{7}$$

For any $\varphi \in H^{-1/2}(\partial\Omega)$, $\partial_\nu \mathcal{S}_{\partial\Omega}[\varphi]$ satisfies the following well-known jump relation (see, for example, [10]):

$$\partial_\nu \mathcal{S}_{\partial\Omega}[\varphi]\big|_{\pm}(x) = \left(\pm \frac{1}{2}I + \mathcal{K}_{\partial\Omega}^* \right)[\varphi](x), \quad x \in \partial\Omega, \tag{8}$$

where the operator $\mathcal{K}_{\partial\Omega}$ is defined by

$$\mathcal{K}_{\partial\Omega}[\varphi](x) = \frac{1}{\omega_d} \int_{\partial\Omega} \frac{\langle y - x, \nu_y \rangle}{|x - y|^d} \varphi(y)\, d\sigma(y) \, , \quad x \in \partial\Omega, \tag{9}$$

and $\mathcal{K}_{\partial\Omega}^*$ is its L^2-adjoint, that is,

$$\mathcal{K}_{\partial\Omega}^*[\varphi](x) = \frac{1}{\omega_d} \int_{\partial\Omega} \frac{\langle x - y, \nu_x \rangle}{|x - y|^d} \varphi(y)\, d\sigma(y) \, , \quad x \in \partial\Omega. \tag{10}$$

Here, $\omega_2 = 2\pi$ and $\omega_3 = 4\pi$. The operator $\mathcal{K}_{\partial\Omega}$ (or $\mathcal{K}_{\partial\Omega}^*$) is called the NP operator on $\partial\Omega$. The operator $\mathcal{K}_{\partial\Omega}$ is also commonly called the double layer potential.

In view of (8), the relation (7) can be transformed to the integral equation

$$\left(\mu_\delta I - \mathcal{K}_{\partial\Omega}^* \right)[\varphi_\delta] = \partial_\nu h \quad \text{on } \partial\Omega, \tag{11}$$

where

$$\mu_\delta := \frac{\epsilon_c + 1}{2(\epsilon_c - 1)} = \frac{k + 1 + i\delta}{2(k - 1) + 2i\delta}. \tag{12}$$

We will come back to the integral equation (11) after recalling some important spectral properties of the NP operator.

The most important spectral property of the operator $\mathcal{K}^*_{\partial\Omega}$ is that it can be realized as a self-adjoint operator on the space $H^{-1/2}(\partial\Omega)$ by introducing a new inner product on it. Let $\langle\,,\,\rangle$ be the usual $H^{-1/2} - H^{1/2}$ duality pairing. We define $\langle\varphi, \psi\rangle_*$ for $\varphi, \psi \in H^{-1/2}(\partial\Omega)$ by

$$\langle\varphi, \psi\rangle_* := -\langle\varphi, \mathcal{S}_{\partial\Omega}[\psi]\rangle. \tag{13}$$

Note that $-\mathcal{S}_{\partial\Omega}$ is a non-negative operator. Since $\mathcal{S}_{\partial\Omega}$ maps $H^{-1/2}(\partial\Omega)$ into $H^{1/2}(\partial\Omega)$, $\langle\varphi, \mathcal{S}_{\partial\Omega}[\psi]\rangle$ is well-defined. The bilinear form $\langle\,,\,\rangle_*$ is actually an inner product on $H_0^{-1/2}(\partial\Omega)$ in two dimensions, and on $H^{-1/2}(\partial\Omega)$ in three dimensions. The NP operator $\mathcal{K}^*_{\partial\Omega}$ is self-adjoint on $H^{-1/2}(\partial\Omega)$ with respect to this inner product. In fact, thanks to the relation

$$\mathcal{S}_{\partial\Omega}\mathcal{K}^*_{\partial\Omega} = \mathcal{K}_{\partial\Omega}\mathcal{S}_{\partial\Omega},$$

which is known as the Plemelj's symmetrization principle (also known as Calderón's identity), we have

$$\langle\varphi, \mathcal{K}^*_{\partial\Omega}[\psi]\rangle_* = -\langle\varphi, \mathcal{S}_{\partial\Omega}\mathcal{K}^*_{\partial\Omega}[\psi]\rangle = -\langle\varphi, \mathcal{K}_{\partial\Omega}\mathcal{S}_{\partial\Omega}[\psi]\rangle = \langle\mathcal{K}^*_{\partial\Omega}[\varphi], \psi\rangle_*.$$

For proofs of properties of the NP operator reviewed so far we refer to the survey paper [22] and references therein.

Spectrum of $\mathcal{K}^*_{\partial\Omega}$ on $H^{-1/2}(\partial\Omega)$, which is denoted by $\sigma(\mathcal{K}^*_{\partial\Omega})$, lies in the interval $(-\frac{1}{2}, \frac{1}{2}]$. Since $\mathcal{K}^*_{\partial\Omega}$ is self-adjoint, $\sigma(\mathcal{K}^*_{\partial\Omega})$ consists of essential spectrum and pure point spectrum, and essential spectrum consists of absolutely continuous spectrum, singularly continuous spectrum, and limit points of eigenvalues (some of them can be void), namely,

$$\sigma(\mathcal{K}^*_{\partial\Omega}) = \sigma_{ess} \cup \sigma_{pp} = \sigma_{ac} \cup \sigma_{sc} \cup \overline{\sigma_{pp}} \tag{14}$$

(see [69]). If $\partial\Omega$ is $C^{1,\alpha}$ for some $\alpha > 0$, then $\mathcal{K}^*_{\partial\Omega}$ is a compact operator on $H^{-1/2}(\partial\Omega)$. In fact, because of orthogonality of the normal vector and the surface or the curve where the operator is defined, we have

$$\frac{|\langle x - y, \nu_x\rangle|}{|x - y|^d} \leqslant \frac{C}{|x - y|^{d-1-\alpha}}, \quad x, y \in \partial\Omega. \tag{15}$$

So the singularity of the integral kernel of $\mathcal{K}^*_{\partial\Omega}$ is weaker than the critical singularity whose order is the same as the dimension of $\partial\Omega$ where the integral is defined. Because of this, $\mathcal{K}^*_{\partial\Omega}$ becomes a compact operator and $\sigma(\mathcal{K}^*_{\partial\Omega})$ consists of eigenvalues of finite multiplicities (except 0 which can have an infinite multiplicity if it is an eigenvalue) accumulating to 0.

Since $\mathcal{K}^*_{\partial\Omega}$ is self-adjoint, the spectral resolution theorem holds [73], namely, there is a family of projection operators $\mathcal{E}(t)$ on $H^{-1/2}(\partial\Omega)$, called the resolution of identity, such that

$$\mathcal{K}^*_{\partial\Omega} = \int_{-1/2}^{1/2} t\, d\mathcal{E}(t). \tag{16}$$

This formula implies, in particular,

$$(\lambda I - \mathcal{K}_{\partial\Omega}^*)^{-1} = \int_{-1/2}^{1/2} \frac{1}{\lambda - t}\, d\mathcal{E}(t). \tag{17}$$

If $\partial\Omega$ is $\mathcal{C}^{1,\alpha}$, (16) takes the form

$$\mathcal{K}_{\partial\Omega}^* = \sum_{j=1}^{\infty} \lambda_j \psi_j \otimes \psi_j, \tag{18}$$

where $\lambda_1, \lambda_2, \ldots$ $(|\lambda_1| \geqslant |\lambda_2| \geqslant \ldots)$ are eigenvalues of $\mathcal{K}_{\partial\Omega}^*$ counting multiplicities, and $\psi_1, \psi_2, \ldots$ are the corresponding (normalized) eigenfunctions.

The following addition formula is proved in [16]: if $\partial\Omega$ is $\mathcal{C}^{1,\alpha}$, then for $x \in \overline{\Omega}$ and $z \in \mathbb{R}^d \setminus \overline{\Omega}$

$$\Gamma(x - z) = -\sum_{j=1}^{\infty} \mathcal{S}_{\partial\Omega}[\psi_j](z)\mathcal{S}_{\partial\Omega}[\psi_j](x) + \mathcal{S}_{\partial\Omega}[\varphi_0](z). \tag{19}$$

If Ω is a ball, then $\mathcal{S}_{\partial\Omega}[\psi_j]$ is a spherical harmonics. Therefore (19) is the expansion of $\Gamma(x - z)$ in terms of the spherical harmonics, which is well-known (see, for example, [51]). If Ω is an ellipsoid, then $\mathcal{S}_{\partial\Omega}[\psi_j]$ is an ellipsoidal harmonic (see [30, Sect. 7.2]) and the formula (19) is due to Heine [35] (it is called the Heine expansion formula in [30]). It is interesting to generalize, if possible, the addition formula (19) to domains with corners. For that (16) may be useful.

We now move back to and look into the integral equation (11). Note that μ_δ converges to $\mu_0 = \frac{k+1}{2(k-1)}$ as $\delta \to 0$. If k is positive, then μ_0 does not belong to $[-1/2, 1/2]$, the interval where $\sigma(\mathcal{K}_{\partial\Omega}^*)$ is contained, and hence the problem (11) is uniquely solvable. Note that the problem (2) is elliptic if k is positive. However, if Ω is a meta-material with the negative k, then $\mu_0 \in (-1/2, 1/2)$, so it is possible (depending on k) for μ_0 to belong to $\sigma(\mathcal{K}_{\partial\Omega}^*)$ and for the solution φ_δ to (11) to blow up as $\delta \to 0$. In fact, according to (18), the solution φ_δ to (11) (assuming that $\partial\Omega$ is $\mathcal{C}^{1,\alpha}$) is given by

$$\varphi_\delta = \sum_{j=1}^{\infty} \frac{\langle \partial_\nu h, \psi_j \rangle_*}{\mu_\delta - \lambda_j} \psi_j. \tag{20}$$

If $\mu_0 = \lambda_i$ for some i and $\langle \partial_\nu h, \psi_i \rangle_* \neq 0$, namely $\partial_\nu h$ has the eigen-mode of ψ_i, then φ_δ blow up as $\delta \to 0$, and so does the solution to (2). This is the plasmon resonance in the quasi-static limit which is one of major reasons for renewed interest in the spectral properties of the NP operator in recent years.

We refer to [13, 14] for the connection of NP spectrum with the sub-wavelength imaging. There asymptotic formulas of resonance frequencies near NP eigenvalues for the Helmholtz equation and the Maxwell system are derived. Convergence of a resonance frequency to a NP eigenvalue for the Helmholtz equation is also proved in [18]. If μ_0 is an eigenvalue of $\mathcal{K}^*_{\partial\Omega}$, then the corresponding k is called a plasmonic eigenvalue and the single layer potential of the corresponding eigenfunction is called a localized surface plasmon [33]. The formula (19) shows that $\mathcal{S}_{\partial\Omega}[\psi_j](z) \to 0$ as $j \to \infty$ for all $z \notin \partial\Omega$ (see also Sect. 3). So, if j is large, then $\mathcal{S}_{\partial\Omega}[\psi_j]$ is localized near $\partial\Omega$. It explains why $\mathcal{S}_{\partial\Omega}[\psi_j]$ is called a *localized surface* plasmon.

3 Analysis of Cloaking by Anomalous Localized Resonance

Suppose that $\partial\Omega$ is $C^{1,\alpha}$ for some $\alpha > 0$ so that $\sigma(\mathcal{K}^*_{\partial\Omega})$ consists of eigenvalues accumulating to 0. Consider the operator-valued function

$$\lambda \mapsto (\lambda I - \mathcal{K}^*_{\partial\Omega})^{-1}.$$

It is a meromorphic function in λ except at $\lambda = 0$. Each NP eigenvalue (other than 0 if 0 is an eigenvalue) is a pole of the function, and 0 is an essential singularity as the limit point of pole. If the dielectric constant k of Ω is -1, then μ_δ in (12) tends to 0, namely, the essential singularity. Near an essential singularity of a meromorphic function, many strange phenomena may occur. It is no exception here. If $k = -1$, then CALR (cloaking by anomalous localized resonance) occurs. We review some recent results on CALR in this section.

Let ϵ be the coefficient given by (1). We consider the following inhomogeneous problem: for a given function f compactly supported in $\mathbb{R}^d \setminus \overline{\Omega}$

$$\begin{cases} \nabla \cdot \epsilon\nabla u = f \text{ in } \mathbb{R}^d, \\ u(x) \to 0 \quad \text{as } |x| \to \infty. \end{cases} \tag{21}$$

The source function f satisfies the condition

$$\int_{\mathbb{R}^d} f\,dx = 0,$$

which is the compatibility condition for existence of the solution to (21). Typically, a dipole is chosen as a source function, that is, $f(x) = a \cdot \nabla\delta_z(x)$, where a is a constant vector, δ_z is the Dirac delta function at z. The location z is assumed to lie outside Ω.

Let u_δ be the solution to the problem (21) and let

$$E_\delta := \Im \int_{\mathbb{R}^d} \epsilon_\delta |\nabla u_\delta|^2\,dx = \int_{\Omega \setminus D} \delta |\nabla u_\delta|^2\,dx \tag{22}$$

($\Im$ for the imaginary part). The problem of CALR is formulated as that of identifying the sources f such that

$$E_\delta \to \infty \quad \text{as } \delta \to 0, \tag{23}$$

and u_δ is bounded outside some radius a.

The quantity E_δ approximately represents the time averaged electromagnetic power produced by the source dissipated into heat. So, (23) implies an infinite amount of energy dissipated per unit time in the limit $\delta \to 0$ which is unphysical. If we scale the source f by a factor of $1/\sqrt{E_\delta}$, then $u_\delta/\sqrt{E_\delta}$ approaches zero outside the radius a. Hence, CALR occurs: the normalized source is essentially invisible from the outside.

The phenomena of anomalous resonance was first discovered in [65] and is related to invisibility cloaking in [62] when Ω is an annulus and $f = a \cdot \nabla \delta_z$. We refer to the recent survey paper [61] for physics related to CALR including superlensing and for a comprehensive list of relevant references. In this paper we discuss the NP spectral nature of CALR.

Let F be the Newtonian potential of f, i.e.,

$$F(x) = \int_{\mathbb{R}^d} \Gamma(x - y) f(y) dy, \quad x \in \mathbb{R}^d.$$

Then, the solution u_δ to (21) takes the form (6) with h replaced with F. There, the potential φ_δ is the solution to the integral equation (11) with $\partial_\nu h$ replaced with $\partial_\nu F$, and hence it is given analogously to (20) by

$$\varphi_\delta = \sum_{j=1}^\infty \frac{\langle \partial_\nu F, \psi_j \rangle_*}{\mu_\delta - \lambda_j} \psi_j.$$

One can see that $\mu_\delta \sim \delta$ since $k = -1$. Here and throughout this paper, we write $A \lesssim B$ to imply that there is a constant C independent of the parameter (in this case it is δ). The meaning of $A \gtrsim B$ is analogous, and $A \sim B$ means both $A \lesssim B$ and $A \gtrsim B$ hold. Suppose that 0 is not an NP eigenvalue on Ω. The CALR condition (23) is equivalent to

$$\widetilde{E}_\delta := \delta \sum_{n=1}^\infty \frac{\langle \partial_\nu F, \psi_n \rangle_*^2}{\delta^2 + \lambda_n^2} \to \infty \quad \text{as } \delta \to 0. \tag{24}$$

In this way, CALR is related to the spectral theory of the NP operator and the result of [62] on CALR with the dipole source on an annulus is extended to more general source f using the spectral approach [5]. Note that the circle has 0 as its NP eigenvalue. By adding another circle the NP eigenvalues are perturbed and 0 is not an NP eigenvalue but the limit point of eigenvalues on an annulus. Furthermore, NP eigenvalues and corresponding eigenfunctions on annuli are explicitly known so that $\widetilde{E}_\delta$ can be computed. This is an advantage of working with annuli. Actually 0 is the only NP eigenvalue, other than $1/2$ which is of multiplicity 1, on a circle. There are planar domains other than disks where 0 is an NP eigenvalue of infinite multiplicity [52]. It is known that if the NP operator on a planar domain is of finite rank so that the eigenvalue 0 has the finite co-multiplicity, then the domain must be a disk [71].

When $f = a \cdot \nabla \delta_z$ for some $z \in \mathbb{R}^d \setminus \overline{\Omega}$ and Ω is an annulus, it is proved in [62] (see also [5]) that there is a virtual radius r_* such that if $|z| < r_*$, then (24) holds and

CALR takes place; if $|z| > r_*$, then E_δ is bounded regardless of δ. In fact, it is shown that r_* is given by

$$r_* = \sqrt{r_e^3/r_i},$$

where r_e and r_i are outer and inner radii of the annulus, respectively. This result has been extended to confocal ellipses in [28]. Since CALR is a phenomenon occurring at the limit point of NP eigenvalues, the structure does not have to be doubly connected. In fact, it is proved in [16] that if Ω is an ellipse, then there is an ellipse Ω_* confocal to Ω such that if $z \in \overline{\Omega_*} \setminus \overline{\Omega}$, then CALR takes place, and it does not if $z \notin \overline{\Omega_*}$ (see Subsect. 3.1). On three dimensional ball or concentric balls CALR does not occur [7, 16]; on concentric balls with folded geometry CALR may occur [6].

Let us look closely the quantity $\widetilde{E}_\delta$ in (24). It is proved in [16] that if $f = a \cdot \nabla \delta_z$, then

$$|\langle \partial_\nu F, \psi_n \rangle_*| \sim |a \cdot \nabla S_{\partial\Omega}[\psi_n](z)|.$$

As a consequence, we have

$$\widetilde{E}_\delta \sim \delta \sum_{n=1}^{\infty} \frac{|a \cdot \nabla S_{\partial\Omega}[\psi_n](z)|^2}{\delta^2 + \lambda_n^2}. \tag{25}$$

Therefore, to investigate the property $\widetilde{E}_\delta \to \infty$, we need to look into the following two questions:

(i) how fast λ_n tends to 0,
(ii) how fast $\nabla S_{\partial\Omega}[\psi_n](z)$ tends to 0,

as $n \to \infty$. The question (i) is about the convergence rate of NP eigenvalues, and the question (ii) is regarding the surface localization of the plasmon $S_{\partial\Omega}[\psi_n]$. Results on the convergence rate of NP eigenvalues are reviewed in [22]. We recall some of them in the following subsection for smooth discussion.

3.1 CALR in Two Dimensions

Let Ω be an ellipse in $\mathbb{R}^2$. The elliptic coordinates $x = (x_1, x_2) = (x_1(\rho, \omega), x_2(\rho, \omega))$, $\rho > 0$ and $0 \leqslant \omega < 2\pi$, is given by

$$x_1(\rho, \omega) = R \cos \omega \cosh \rho, \quad x_2(\rho, \omega) = R \sin \omega \sinh \rho.$$

We denote the elliptic coordinates of x by $\rho = \rho_x$ and $\omega = \omega_x$. Then $\partial\Omega$ is represented by

$$\partial\Omega = \{x \in \mathbb{R}^2 : \rho_x = \rho_0\} \tag{26}$$

for some $\rho_0 > 0$. The number ρ_0 is called the elliptic radius of Ω.

The NP eigenvalues on $\partial\Omega$ are

$$\pm\lambda_n = \pm\frac{1}{2e^{2n\rho_0}}, \quad n = 1, 2, \ldots,$$

and eigenfunctions corresponding $+\lambda_n$ and $-\lambda_n$ are respectively given by

$$\psi_n^c(\omega) := \Xi(\rho_0, \omega)^{-1} \cos n\omega, \quad \psi_n^s(\omega) := \Xi(\rho_0, \omega)^{-1} \sin n\omega,$$

where

$$\Xi = \Xi(\rho_0, \omega) := R\sqrt{\sinh^2 \rho_0 + \sin^2 \omega}$$

(see, for example, [28]). Using these facts, it is proved in [16] that

$$\widetilde{E}_\delta \sim \begin{cases} \delta^{-2+\rho_z/\rho_0}|\log\delta| & \text{if } \rho_0 < \rho_z < 3\rho_0, \\ \delta|\log\delta|^2 & \text{if } \rho_z = 3\rho_0, \\ \delta & \text{if } \rho_z > 3\rho_0, \end{cases}$$

as $\delta \to 0$, where z is the position of the source. As a consequence, the following theorem is obtained.

Theorem 3.1 *Let $\partial\Omega$ be the ellipse given by (26). CALR takes place if $\rho_0 < \rho_z \leqslant 2\rho_0$ and does not take place if $\rho_z > 2\rho_0$, namely, the critical (elliptic) radius for CALR is $2\rho_0$.*

If Ω is a planar domain with the real analytic boundary, then it is not known yet if CALR takes place. One may attempt to prove the following problem:

Problem 1. Is it true that if the location z of the source is close enough to $\partial\Omega$, then CALR takes place, and if z is sufficiently away from $\partial\Omega$, it does not take place?

In relation to this, the question (i) above has been answered in [19] (see also [22] and references therein for related work): NP eigenvalues converge to 0 exponentially fast and the rate is determined by the modified maximal Grauert radius of $\partial\Omega$. The modified maximal Grauert radius is basically the radius up to which the real analytic defining function of $\partial\Omega$ is extended as a complex analytic function (see [19] for a precise definition). In fact, if λ_n is the positive NP eigenvalues on $\partial\Omega$ enumerated in descending order, then for any $\epsilon < \epsilon_{\partial\Omega}$ there exists a constant C such that

$$\lambda_n \leqslant Ce^{-\epsilon n}$$

for all n, Therefore, we obtain from (25)

$$\widetilde{E}_\delta \gtrsim \delta \sum_{n=1}^{\infty} \frac{|a \cdot \nabla\mathcal{S}_{\partial\Omega}[\psi_n](z)|^2}{\delta^2 + e^{-2\epsilon n}}$$

for any $\epsilon < \epsilon_{\partial\Omega}$. However, it is not known whether CALR occurs or not because of lack of knowledge on the localization of plasmon in the question (ii). In fact, to the best of author's knowledge, there is no result on (ii) in two dimensions except on disks, annuli, and ellipses where NP eigenvalues and eigenfunctions are known explicitly. This brings the following problem

Problem 2. How fast does $\mathcal{S}_{\partial\Omega}[\psi_n](z)$ tend to 0 as $n \to \infty$ when $\partial\Omega$ is real analytic and of general shape?

3.2   CALR in Three Dimensions

Let Ω be a bounded domain in $\mathbb{R}^3$ with the smooth boundary. It is proved in [63] (see also [22]) that $\{\lambda_n\}_{n\in\mathbb{N}}$ asymptotically behaves like

$$\lambda_n^2 \sim C_{\partial\Omega} n^{-1} \quad \text{as } n \to \infty, \tag{27}$$

in the sense that $\lambda_n^2 n \to C_{\partial\Omega}$ as $n \to \infty$. The constant $C_{\partial\Omega}$ is given by

$$C_{\partial\Omega} = \frac{3W(\partial\Omega) - 2\pi\chi(\partial\Omega)}{128\pi},$$

where $W(\partial\Omega)$ and $\chi(\partial\Omega)$, respectively, are the Willmore energy and the Euler characteristic of the boundary surface $\partial\Omega$. The Willmore energy is defined by

$$W(\partial\Omega) = \int_{\partial\Omega} H(x)^2 dS,$$

where $H(x)$ is the mean curvature and it is known that $W(\partial\Omega) \geqslant 4\pi$ [59] (see also [22]). Thus we have $C_{\partial\Omega} > 0$, and hence

$$\widetilde{E}_\delta \sim \delta \sum_{n=1}^{\infty} \frac{|a \cdot \nabla\mathcal{S}_{\partial\Omega}[\psi_n](z)|^2}{\delta^2 + n^{-1}}. \tag{28}$$

The following theorem is proved in [21].

Theorem 3.2 *Let Ω be a strictly convex bounded domain in $\mathbb{R}^3$ with the C^∞-smooth boundary. For any compact set K in $\mathbb{R}^3 \setminus \overline{\Omega}$ and for each integers k and s there is a constant $C_{k,s}$ such that*
$$\|\mathcal{S}_{\partial\Omega}[\psi_n]\|_{C^k(K)} \leq C_{k,s} n^{-s}$$

for all sufficiently large n.

The main ingredient in proving this theorem is the fact that since $\partial\Omega$ is C^∞-smooth, the NP operator $\mathcal{K}_{\partial\Omega}^*$ is known to be a strictly homogeneous pseudo-differential operator of order -1 [32], and its principal symbol is positive definite

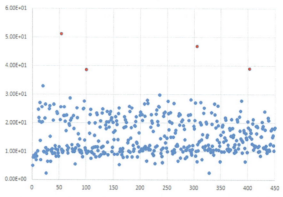

Fig. 1 The graph of $\|\mathcal{S}_{\partial\Omega}[\psi_n]\|_{L^2(X)}$ ($1 \leqslant n \leqslant 450$) after some normalization on the Clifford torus. The horizontal axis represents positive eigenvalues of the NP operator enumerated in decreasing order up to 450. The red dots indicate values drastically larger than neighboring points. They occur at 53rd, 100th, 305th and 402nd eigenvalues. (The figure is from [21]. The region X where $\mathcal{S}_{\partial\Omega}[\psi_n]$ is computed is the rectangle-shaped cross section shown in Fig. 2 and $\mathcal{S}_{\partial\Omega}[\psi_n]$ is normalized.)

if Ω is strictly convex as proved in [63, 64]. Thus all eigenvalues of $\mathcal{K}^*_{\partial\Omega}$, except possibly finitely many, are positive [64]. By altering $\mathcal{K}^*_{\partial\Omega}$ on a finite dimensional subspace if necessary, we can realize $\mathcal{K}^*_{\partial\Omega}$ as a positive definite pseudo-differential operator of order -1. This fact together with (27) leads us to Theorem 3.2.

We infer from Theorem 3.2 and (28) that $\widetilde{E}_\delta < +\infty$, and hence CALR does not take place on three-dimensional strictly convex bounded domains.

In order to see what happens if $\partial\Omega$ is not convex, the first 450 plasmons, namely, $\mathcal{S}_{\partial\Omega}[\psi_n]$ ($1 \leqslant n \leqslant 450$), corresponding to largest eigenvalues in descending order, are computed numerically on a cross section close to $\partial\Omega$ when Ω is the Clifford torus in [21]. To our surprise, there are four out of 450 plasmons which do not decay fast enough (see Fig. 1). It is utterly interesting to investigate it rigorously and explore the possible connection to CALR. We include in this paper the figures from [21] to compare $\mathcal{S}_{\partial\Omega}[\psi_n]$ of fast decay and slow decay (see Fig. 2). We also include figures of two exceptional eigenfunctions together with one non-exceptional one for comparison. The exceptional eigenfunctions have a very interesting feature which we discuss in the next section (see Fig. 3).

For general domains the following theorem is proved using the addition formula (19) (we only state a special case of the theorem obtained in [21]).

Theorem 3.3 *Suppose that Ω is a bounded domain in $\mathbb{R}^3$ with the $C^{1,\alpha}$ smooth boundary for some $\alpha > 0$. For any compact set K in $\mathbb{R}^3 \setminus \overline{\Omega}$,*

$$\|\mathcal{S}_{\partial\Omega}[\psi_n]\|_{C^1(K)} = o(n^{-1/2}) \quad \text{almost surely as } n \to \infty.$$

For a sequence $\{a_n\}$ of numbers and a non-negative number s, we say $a_n = o(n^{-s})$ almost surely as $n \to \infty$ if

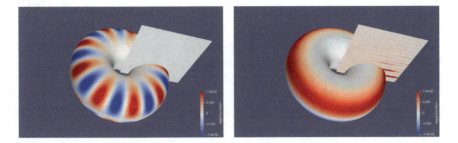

Fig. 2 $\mathcal{S}_{\partial\Omega}[\psi_n]$ of fast decay (left) and of slow decay (right). The rectangular cross section represents the region where $\mathcal{S}_{\partial\Omega}[\psi_n]$ is evaluated, and the color on it represents its value. (Figures are from [21].)

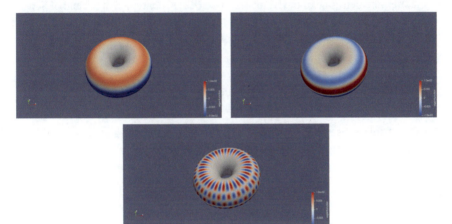

Fig. 3 Top: exceptional eigenfunctions whose single layer potentials do not decay fast. They do not oscillate in the toroidal direction. Bottom: a non-exceptional eigenfunction whose single layer potential decays fast. It oscillates in the toroidal direction. (Figures are from [21].)

$$\lim_{\delta\downarrow 0}\limsup_{N\to\infty}\frac{\sharp\{n < N \ : \ |a_n| > \delta n^{-s}\}}{N} = 0.$$

It is equivalent to existence of a subsequence $\{n_k\}$ such that $a_{n_k} = o(n_k^{-s})$ as $k \to \infty$ and $\lim_{k\to\infty} n_k/k = 1$.

4 Concavity and Negative Eigenvalues

In two dimensions the NP spectrum always appears in pairs, namely, if $\lambda \in \sigma(\mathcal{K}^*_{\partial\Omega})$, then $-\lambda \in \sigma(\mathcal{K}^*_{\partial\Omega})$. This can be proved using existence of harmonic conjugates. However, there are domains in three dimensions where the NP operators have only positive eigenvalues: the NP eigenvalues on a sphere are $1/(4n + 2)$ for $n = 0, 1, 2 \ldots$, and they are all positive on prolate spheroids [3]. The first example of three-dimensional domains with a negative NP eigenvalue was found in [2]: it is an oblate spheroid. We now know that concavity allows negative NP eigenvalues which we review in this section.

Let us first recall the discussion in [17] on possible advantage of having negative eigenvalues. Suppose that $\epsilon_c = k + i\delta$ is the dielectric constant of Ω as before and assume $\delta = 0$ so that $\epsilon_c = k$. Then plasmon resonance in the quasi-static limit occurs if

$$\frac{k + 1}{2(k - 1)} = \lambda, \tag{29}$$

where λ is an eigenvalue of the NP operator on $\partial\Omega$ as mentioned at the end of Sect. 2. The relation (29) can be achieved by a larger k (a smaller $|k|$) if λ is negative (see Fig. 4). This may yield an advantage in practice.

4.1 A Concavity Condition for Negative Eigenvalues

Results of this subsection are from [38]. For a fixed $r > 0$ and $p \in \mathbb{R}^3$, let $T_p : \mathbb{R}^3 \setminus \{p\} \to \mathbb{R}^3 \setminus \{p\}$ be the inversion in a sphere, namely,

$$T_p x := \frac{r^2}{|x - p|^2}(x - p) + p. \tag{30}$$

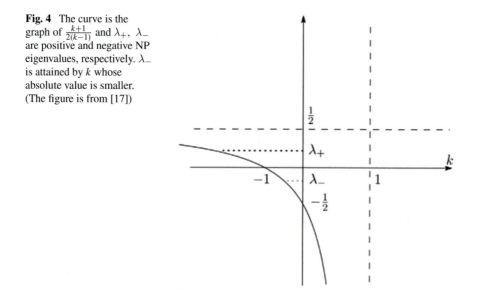

Fig. 4 The curve is the graph of $\frac{k+1}{2(k-1)}$ and λ_+, λ_- are positive and negative NP eigenvalues, respectively. λ_- is attained by k whose absolute value is smaller. (The figure is from [17])

For a given bounded domain Ω in $\mathbb{R}^3$, let $\partial\Omega_p^*$ be the inversion of $\partial\Omega$, i.e., $\partial\Omega_p^* = T_p(\partial\Omega)$.

The following theorem is obtained in [38].

Theorem 4.1 *Let Ω be a bounded domain in $\mathbb{R}^3$ whose boundary is Lipschitz continuous. If there are $p \in \Omega$ and $x \in \partial\Omega$ such that*

$$(x - p) \cdot \nu_x < 0 \tag{31}$$

and $\partial\Omega$ is C^1 near x, then either $\sigma(\mathcal{K}_{\partial\Omega}^*)$ or $\sigma(\mathcal{K}_{\partial\Omega_p^*}^*)$ has a negative value.

The condition (31) indicates that $\partial\Omega$ is concave with respect to $p \in \Omega$. For example, this condition is fulfilled if there is a point on $\partial\Omega$ where the Gaussian curvature is negative. Thus the following corollary is obtained.

Corollary 4.2 *Suppose $\partial\Omega$ is C^2 smooth. If there is a point on $\partial\Omega$ where the Gaussian curvature is negative, then either $\mathcal{K}_{\partial\Omega}^*$ or $\mathcal{K}_{\partial\Omega_p^*}^*$ for some $p \in \Omega$ has a negative eigenvalue.*

Let us briefly see how Theorem 4.1 is proved. Just for simplicity we assume $p = 0$ and denote $\partial\Omega_p^*$ by $\partial\Omega^*$. For a function φ defined on $\partial\Omega$, define φ^* on $\partial\Omega^*$ by

$$\varphi^*(x^*) := \varphi(x)\frac{|x|^d}{r^d}, \quad x^* = T_p x.$$

The following identity holds for all φ:

$$\langle \mathcal{K}_{\partial\Omega^*}^*[\varphi^*], \varphi^* \rangle_{\partial\Omega^*} + \langle \mathcal{K}_{\partial\Omega}^*[\varphi], \varphi \rangle_{\partial\Omega} = \int_{\partial\Omega} \frac{y \cdot \nu_y}{|y|^2} |\mathcal{S}_{\partial\Omega}[\varphi](y)|^2 \, dS. \tag{32}$$

Here, $\langle \, , \, \rangle_{\partial\Omega^*}$ and $\langle \, , \, \varphi \rangle_{\partial\Omega}$ denote the inner product (13) on $\partial\Omega^*$ and $\partial\Omega$, respectively. In fact, (32) is derived using the following three transformation formulas,

$$dS(x^*) = \frac{r^4}{|x|^4} dS(x)$$

for the surface measure dS,

$$\Gamma(x^* - y^*) = \frac{|x||y|}{r^2} \Gamma(x - y)$$

for the fundamental solution, and

$$\nu_{x^*} = (-1)^m \left(I - 2\frac{x}{|x|}\frac{x^t}{|x|} \right)\nu_x$$

for the normal vectors.

If (31) holds, namely, $x \cdot \nu_x < 0$ for some $x \in \partial\Omega$, then, since $\partial\Omega$ is C^1 near x and $\mathcal{S}_{\partial\Omega} : H^{-1/2}(\partial\Omega) \to H^{1/2}(\partial\Omega)$ is invertible, we can choose $\varphi \in H^{-1/2}(\partial\Omega)$ so that $\mathcal{S}_{\partial\Omega}[\varphi]$ is supported in a small neighborhood of x and

$$\int_{\partial\Omega} \frac{y \cdot \nu_y}{|y|^2} |\mathcal{S}_{\partial\Omega}[\varphi](y)|^2 \, dS < 0.$$

Thus we have

$$\langle \mathcal{K}^*_{\partial\Omega^*}[\varphi^*], \varphi^* \rangle_{\partial\Omega^*} + \langle \mathcal{K}^*_{\partial\Omega}[\varphi], \varphi \rangle_{\partial\Omega} < 0. \tag{33}$$

Thus, the numerical range of either $\mathcal{K}^*_{\partial\Omega^*}$ or $\mathcal{K}^*_{\partial\Omega}$ has a negative element. The numerical range of the operator T is defined to be

$$W(T) := \{ \langle Tx, x \rangle \; ; \; \|x\| = 1 \}$$

and convex (Hausdorff-Toeplitz theorem). In particular, if T is self-adjoint, then $W(T)$ is an interval and its endpoints are eigenvalues of T (see [34]). Thus Theorem 4.1 follows from (33).

4.2 NP Spectrum on Tori

We now review the results in [15] on NP spectrum on tori. We do so in some detail having in mind the possible connection to CALR as mentioned in the previous section.

The toroidal coordinate system (ξ, θ, η) for the Cartesian coordinates (x_1, x_2, x_3) is given by

$$x_1 = \frac{R_0\sqrt{1 - \xi^2} \cos\eta}{1 - \xi \cos\theta}, \quad x_2 = \frac{R_0\sqrt{1 - \xi^2} \sin\eta}{1 - \xi \cos\theta}, \quad x_3 = -\frac{R_0\xi \sin\theta}{1 - \xi \cos\theta}.$$

Here, $R_0 = \sqrt{r_0^2 - a^2}$ where r_0 and a are the major and minor radii, respectively, of a toroidal system. The variable ξ $(0 < \xi < 1)$ is similar to the minor radius, θ $(0 \le \theta < 2\pi)$ is the poloidal angle, and η $(0 \le \eta < 2\pi)$ is the toroidal angle (see [25] for the toroidal coordinate system).

The surface $\xi = $ constant is a torus, on which (θ, η) is the coordinate system. Let, with the fixed ξ,

$$\mu(\eta - \eta') := \frac{1}{\xi^2} + \left(1 - \frac{1}{\xi^2}\right) \cos(\eta - \eta')$$

and

$$\chi(\theta) := 1 - \xi \cos\theta.$$

Then, $\mathcal{K}^*_{\partial\Omega}[\varphi]$ for a function φ on the torus can be written as

$$\mathcal{K}^*_{\partial\Omega}[\varphi](\theta, \eta) = \int_0^{2\pi} \int_0^{2\pi} k(\theta, \theta'; \eta - \eta')\varphi(\theta', \eta')\, d\theta'\, d\eta',$$

where

$$k(\theta, \theta'; \eta - \eta') = \frac{1 - \xi^2}{8\pi\sqrt{2}\xi} \frac{\chi(\theta)^{1/2}}{\chi(\theta')^{3/2}} \frac{1}{(\mu(\eta - \eta') - \cos(\theta - \theta'))^{1/2}}$$

$$- \frac{1 - \xi^2}{8\pi\sqrt{2}\xi^3} \frac{\chi(\theta)^{3/2}}{\chi(\theta')^{3/2}} \frac{1 - \cos(\eta - \eta')}{(\mu(\eta - \eta') - \cos(\theta - \theta'))^{3/2}}.$$

Since $0 < \xi < 1$, χ is a positive smooth function. Thus any function φ in $H^{-1/2}$ on the torus admits the Fourier series expansion

$$\varphi(\theta, \eta) = \chi(\theta)^{3/2} \sum_{k=-\infty}^{\infty} \hat{\varphi}_k(\theta)e^{ik\eta}. \tag{34}$$

We thus have

$$\mathcal{K}^*_{\partial\Omega}[\varphi](\theta, \eta) = C_\xi \chi(\theta)^{3/2} \sum_{k=-\infty}^{\infty} \mathcal{K}_k[\hat{\varphi}_k](\theta)e^{ik\eta}, \tag{35}$$

where

$$C_\xi = \frac{1 - \xi^2}{8\pi\sqrt{2}\xi}$$

and the operator $\mathcal{K}_k$ is defined by

$$\mathcal{K}_k[f](\theta) := \int_0^{2\pi} a_k(\theta, \theta') f(\theta')\, d\theta'$$

with

$$a_k(\theta, \theta') = \chi(\theta)^{-1} \int_0^{2\pi} \frac{e^{-ik\eta'}}{(\mu(\eta') - \cos(\theta - \theta'))^{1/2}}\, d\eta'$$

$$- \frac{1}{\xi^2} \int_0^{2\pi} \frac{(1 - \cos\eta')e^{-ik\eta'}}{(\mu(\eta') - \cos(\theta - \theta'))^{3/2}}\, d\eta'.$$

The single layer potential also admits the decomposition

$$\mathcal{S}_{\partial\Omega}[\varphi](\theta, \eta) = \frac{R_0\sqrt{1 - \xi^2}}{4\pi\sqrt{2}} \chi(\theta)^{1/2} \sum_{k=-\infty}^{\infty} \mathcal{S}_k[\hat{\varphi}_k](\theta)e^{ik\eta}, \tag{36}$$

where

$$\mathcal{S}_k[f](\theta) := \int_0^{2\pi} s_k(\theta - \theta') f(\theta') \, d\theta'$$

with

$$s_k(\theta) := \int_0^{2\pi} \frac{e^{-ik\eta'}}{(\mu(\eta') - \cos\theta)^{1/2}} \, d\eta'.$$

Let $H^{-1/2}(T)$ be the Sobolev space of order $-1/2$ on the unit circle T. If we define $\langle \, , \, \rangle_k$ by

$$\langle f, g \rangle_k := \int_0^{2\pi} f(\theta) \overline{\mathcal{S}_k[g](\theta)} \, d\theta, \quad f, g \in H^{-1/2}(T), \tag{37}$$

then it is an inner product on $H^{-1/2}(T)$ for each k and the following relation holds:

$$\langle \varphi, \psi \rangle_* = \frac{R_0^{\,3} \xi (1 - \xi^2)}{4\pi\sqrt{2}} \sum_{k=-\infty}^{\infty} \langle \hat{\varphi}_k, \hat{\psi}_k \rangle_k, \tag{38}$$

where $\hat{\varphi}_k = \hat{\varphi}_k(\theta)$ is the Fourier coefficient as defined in (34). Moreover, $\mathcal{K}_k$ is compact and self-adjoint on $H^{-1/2}(T)$, that is,

$$\langle \mathcal{K}_k[g_1], g_2 \rangle_k = \langle g_1, \mathcal{K}_k[g_2] \rangle_k.$$

We infer from the relation (35) that if λ is an eigenvalue of $\mathcal{K}_k$ and g is the corresponding eigenfunction, then $C_\xi \lambda$ is an eigenvalue of $\mathcal{K}_{\partial\Omega}^*$ and the corresponding eigenfunction is given by

$$\varphi(\theta, \eta) = \chi(\theta)^{3/2} g(\theta) e^{ik\eta}. \tag{39}$$

Note that the function φ oscillates in toroidal direction if $k \neq 0$. As we saw in Fig. 3, the exceptional eigenfunctions, whose single layer potentials (namely, plasmon) do not decay fast, do not oscillate in the toroidal direction. On the other hand, the non-exceptional eigenfunctions oscillate in the toroidal direction.

It is proved using the stationary phase method that for any $0 < \xi < 1$ there exists a positive integer k_0 such that the numerical range of $\mathcal{K}_k$ has both positive and negative values and hence $\mathcal{K}_k$ has both positive and negative eigenvalues for all $k \in \mathbb{Z}$ with $|k| > k_0$. Thus we have the following theorem from Hausdorff-Toeplitz theorem again:

Theorem 4.3 *The NP operator on tori has infinitely many negative eigenvalues.*

Any bounded domain with $C^{1,\alpha}$ boundary has infinitely many positive NP eigenvalues.

Two questions arise naturally:

Problem 3. The exceptional eigenvalues (those NP eigenvalues such that the single layer potentials of the corresponding eigenfunctions decay slowly as discussed in Sect. 3) are from $k = 0$, namely, $C_\xi \lambda$ where λ is an eigenvalue of $\mathcal{K}_0$?

Problem 4. Is it true that all eigenfunctions of $\mathcal{K}^*_{\partial\Omega}$ are of the form (39), in particular, any eigenvalue of $\mathcal{K}^*_{\partial\Omega}$ is C_ξ times an eigenvalue of $\mathcal{K}_k$ for some k?

4.3 Further Results

We mention general results in [64]. For more precise statements of the results and discussions on them, we refer to [22].

Theorem 4.4 *Let $\Omega \subset \mathbb{R}^3$ be a bounded domain with C^∞ boundary.*

(i) *If one of the principal curvatures at $x \in \partial\Omega$ is positive, then there are infinitely many negative NP eigenvalues.*
(ii) *If the principal curvatures are negative at every $x \in \partial\Omega$, then there are at most finitely many negative NP eigenvalues.*

In a recent paper [37] the NP spectrum on surfaces of revolution of planar curves with a corner is considered and the continuous spectrum can be both positive or negative depending on the angle of the corner. It is the first attempt to investigate continuous spectrum on three-dimensional domain with corners.

5 NP Spectrum on Polygonal Domains

In order to introduce an outstanding problem on essential spectrum, we begin the discussion in this section by briefly reviewing results on essential spectrum. For more extensive review and interesting historical account, we refer to [22].

Let $\mathcal{E}(t)$ be the resolution of identity given in (16). For each $\varphi \in H^{-1/2}(\partial\Omega)$, the measure $\mu_\varphi := \langle \mathcal{E}(t)\varphi, \varphi \rangle_*$ is called the spectral measure associated with φ. According to Lebesgue decomposition theorem, $H^{-1/2}(\partial\Omega)$ can be decomposed as

$$H^{-1/2}(\partial\Omega) = H_{pp} \oplus H_{ac} \oplus H_{sc},$$

where H_{sc} is the collection of all φ such that μ_φ is singularly continuous, and H_{pp} and H_{ac} are defined likewise. The singularly continuous $\sigma_{sc} = \sigma_{sc}(\mathcal{K}^*_{\partial\Omega})$ is the spectrum of $\mathcal{K}^*_{\partial\Omega}$ when restricted H_{sc}, and σ_{ac} and σ_{pp} are defined likewise. See [69].

Let Ω be a bounded Lipschitz domain in $\mathbb{R}^2$ with finite number of corners whose inner angles are $\alpha_1, \ldots, \alpha_N$. Let

$$b_{ess} = \frac{1}{2} \max_{1 \leqslant j \leqslant N} \left(1 - \frac{\alpha_j}{\pi} \right). \tag{40}$$

It is proved in [67] that b_{ess} is a bound of $\sigma_{ess}(\mathcal{K}^*_{\partial\Omega})$. In [47], a lens domain (an intersection of two disks) is considered and a complete spectral resolution of $\mathcal{K}^*_{\partial\Omega}$ is derived. It enables us to infer that

$$\sigma(\mathcal{K}^*_{\partial\Omega}) = \sigma_{ess}(\mathcal{K}^*_{\partial\Omega}) = \sigma_{ac}(\mathcal{K}^*_{\partial\Omega}) = [-b_{ess}, b_{ess}].$$

In particular, it implies that $\sigma_{pp}(\mathcal{K}^*_{\partial\Omega})$ and $\sigma_{sc}(\mathcal{K}^*_{\partial\Omega})$ are void on a lens domain. If Ω be a bounded Lipschitz domain in $\mathbb{R}^2$ with finite number of corners, it is proved in [68] that $\sigma_{ess}(\mathcal{K}^*_{\partial\Omega}) = [-b_{ess}, b_{ess}]$, and in [66] that $\sigma_{sc}(\mathcal{K}^*_{\partial\Omega})$ is void.

An interesting problem arises:

Problem 5. Find geometric conditions on $\partial\Omega$ which guarantee $\sigma_{sc}(\mathcal{K}^*_{\partial\Omega}) \neq \varnothing$. (Since no example of domains with nonempty singularly continuous spectrum is known, even construction of a single example of such a domain would be interesting.)

We now review the results of [36] where the question whether $\sigma_{pp}(\mathcal{K}^*_{\partial\Omega})$ is void or not. The crux of the matter is to use resonance to distinguish σ_{pp} and σ_{ac}. This idea also appears in [16].

Let $f \in H_0^{-1/2}(\partial\Omega)$. For $t \in (-1/2, 1/2)$ and $\delta > 0$, let $\varphi_{t,\delta}$ be the solution of the integral equation

$$((t + i\delta)I - \mathcal{K}^*_{\partial\Omega})[\varphi_{t,\delta}] = f \quad \text{on } \partial\Omega.$$

By the spectral resolution (17), the solution is given by

$$\varphi_{t,\delta} = \int_{\sigma(\mathcal{K}^*_{\partial\Omega})} \frac{1}{t + i\delta - s} \, d\mathcal{E}(s)[f],$$

and hence

$$\|\varphi_{t,\delta}\|_*^2 = \langle \varphi_{t,\delta}, \varphi_{t,\delta} \rangle_* = \int_{\sigma(\mathcal{K}^*_{\partial\Omega})} \frac{1}{(s - t)^2 + \delta^2} \, d\langle \mathcal{E}(s)[f], f \rangle_*.$$

If $t \notin \sigma(\mathcal{K}^*_{\partial\Omega})$, one can immediately see from (20) that $\|\varphi_{t,\delta}\|_* < C$ for some C regardless of δ.

If $t \in \sigma(\mathcal{K}^*_{\partial\Omega})$, then $\|\varphi_{t,\delta}\|_*$ may blow up as $\delta \to 0$. The key idea is that the blow-up rate at the eigenvalue t is different from that at continuous spectrum. To see this, we recall that an eigenvalue t of $\mathcal{K}^*_{\partial\Omega}$ is characterized by discontinuity $\mathcal{E}(t+) - \mathcal{E}(t) \neq 0$ (and t is isolated) (see [73]). So, if f satisfies

$$\langle \mathcal{E}(t+)[f], f \rangle_* - \langle \mathcal{E}(t)[f], f \rangle_* > 0, \tag{41}$$

then

$$\|\varphi_{t,\delta}\|_*^2 \geq \frac{\langle \mathcal{E}(t+)[f], f \rangle_* - \langle \mathcal{E}(t)[f], f \rangle_*}{\delta^2},$$

and hence

$$\|\varphi_{t,\delta}\|_*^2 \approx \delta^{-2}.$$

If $t \in \sigma_{ac}(\mathcal{K}^*_{\partial\Omega})$, then there is f such that the spectral measure $\langle \mathcal{E}_s[f], f \rangle_*$ is absolutely continuous near t, namely, there is $\epsilon > 0$ and a function $\mu_f(s)$ which is integrable on $[t - \epsilon, t + \epsilon]$ such that

$$d\langle \mathcal{E}(s)[f], f \rangle_* = \mu_f(s)ds, \quad s \in [t - \epsilon, t + \epsilon]. \tag{42}$$

Then it is proved that

$$\lim_{\delta \to 0} \delta \|\varphi_{t,\delta}\|_*^2 = \frac{\pi}{2}(\mu_f(t+) + \mu_f(t-)) > 0,$$

and

$$\lim_{\delta \to 0} \delta^2 \|\varphi_{t,\delta}\|_*^2 = 0.$$

Define an indicator function $\alpha_f(t)$ by

$$\alpha_f(t) := \sup \left\{ \alpha \ \Big| \ \limsup_{\delta \to 0} \delta^\alpha \|\varphi_{t,\delta}\|_* = \infty \right\}, \quad t \in (-1/2, 1/2). \tag{43}$$

We see that $0 \leqslant \alpha_f(t) \leqslant 1$ for all t. The following theorem for classification of NP spectra of is obtained in [36].

Theorem 5.1 *Let* $f \in H_0^{-1/2}(\partial\Omega)$.

(i) *If* $\alpha_f(t) > 0$, *then* $t \in \sigma(\mathcal{K}^*_{\partial\Omega})$.
(ii) *If* $\alpha_f(t) = 1$ *and* t *is isolated, then* $t \in \sigma_{pp}(\mathcal{K}^*_{\partial\Omega})$.
(iii) *If* $1/2 \leqslant \alpha_f(t) < 1$, *then* $t \in \sigma_{ess}(\mathcal{K}^*_{\partial\Omega})$.

The indicator function $\alpha_f(t)$ can be computed numerically using the following identity:

$$\alpha_f(t) = -\lim_{\delta \to 0} \frac{\log \|\varphi_{t,\delta}\|_*}{\log \delta}$$

if the limit exists.

The source function f should satisfy (41) and (42) with $\mu_f(t+) + \mu_f(t-) > 0$. In [36], $f_z(x) = \nu(x) \cdot \nabla(a \cdot \nabla_x \Gamma(x - z))$ (a is a constant vector and z lies outside Ω) is chosen as a source function because for any $\varphi \in H_0^{-1/2}(\partial\Omega)$, $\langle f_z, \varphi \rangle_* \neq 0$ for almost all z among several advantages. Then the indicator function is modified as

$$\alpha_\sharp(t) := \max_{1 \leqslant j \leqslant N} \{\alpha_{f_{z_j}}(t)\}$$

for some $z_1, \ldots, z_N$.

This method of characterizing NP spectra has been tested for various polygonal domains, ellipses perturbed by a corner, and so on. The computational results clearly show effectiveness of the method in distinguishing eigenvalues from continuous spectrum. Some results show that eigenvalues can be embedded inside the continuous spectrum (see Fig. 5). Lately, it is rigorously proved that it is possible that infinitely many eigenvalues are embedded in the continuous spectrum [54, 55].

The computational results on rectangles are particularly interesting. We know by (40) that if Ω is rectangle, then $\sigma_{ess}(\mathcal{K}^*_{\partial\Omega}) = [-1/4, 1/4]$. However, as the aspect ratio of the rectangle gets larger, more and more eigenvalues show up (see Fig. 6). We can formulate an interesting problem out of these numerical experiments.

Problem 6. Is it true that there is a sequence of numbers $1 = r_0 < r_1 < r_2 < \cdots$ such that if the aspect ratio of the rectangle lies in (r_{k-1}, r_k), then the number of positive NP eigenvalues is k? The first number r_1 seems to be around 2.201592 (see the second figure in Fig. 6).

Figure 7 show the NP spectrum of an isosceles triangle. The interval of continuous spectrum is determined by the smallest interior angle as explained earlier. It shows no eigenvalue. It is not clear whether triangles have no NP eigenvalues or not and it would be interesting to clarify this.

Problem 7. Prove or disprove that there is no NP eigenvalues on triangles.

It is also interesting to investigate the NP spectral structure on polyhedra. We mention a recent paper [53] in this regard.

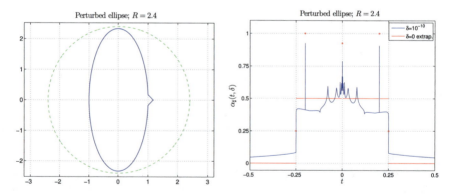

Fig. 5 An ellipse perturbed by a corner. Two eigenvalues are embedded in the continuous spectrum. (Figures from [36])

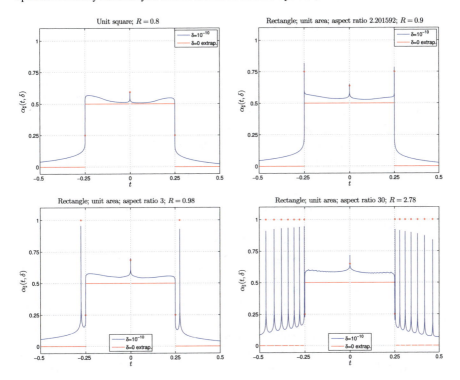

Fig. 6 Graphs of the indicator function $\alpha_\sharp$ on rectangles of aspect ratios, $r = 1$, 2.201592, 3, 30. When $r = 2.201592$, eigenvalues just about to emerge, and more and more eigenvalues emerge as the aspect ratio increases. (Figures from [36])

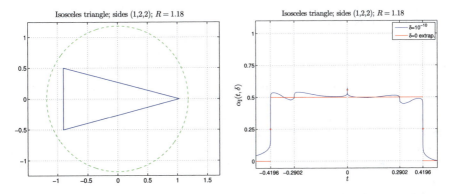

Fig. 7 NP spectrum of the isosceles triangle with sides 1, 2 and 2. There is no eigenvalue. (Figures from [36])

6 Spectral Structure of Thin Domains

Motivated by the numerical study on the NP spectral structure on rectangles as explained in the previous section, the spectral structure of the NP operators on thin domains has been investigated in two dimensions [20] and in three dimensions [17]. We review those results in this section.

There is another motivation behind these works. They are motivated by observations that as the boundary $\partial\Omega$ of the domain becomes singular in some sense, the corresponding NP spectrum seems to approach to the bound $\pm 1/2$. For example, as we saw at the beginning of the previous section, if a planar domain has corners and if a corner gets sharper and the domain becomes needle-like around the corner, then the essential spectrum approaches $[-1/2, 1/2]$. If Ω consists of two strictly convex planar domains and boundaries get closer, then more and more eigenvalues of the corresponding NP operator approach $\pm 1/2$ [26, 27]. This causes stress concentration in the narrow region between two inclusions (see Sect. 7). Results to be reviewed in this section show that as the domain gets thinner, NP spectrum approaches $[-1/2, 1/2]$ in two dimensions. In three dimensions, it approaches either $[-1/2, 1/2]$ or $[0, 1/2]$ depending upon the kind of thinness (thin and long, or thin and flat). So, in some case there is no negative eigenvalue.

In this section we work with $\mathcal{K}_{\partial\Omega}$, not $\mathcal{K}^*_{\partial\Omega}$, since it is more convenient. The spectrum of $\mathcal{K}_{\partial\Omega}$ on $H^{1/2}(\partial\Omega)$ is the same as the spectrum of $\mathcal{K}^*_{\partial\Omega}$ on $H^{-1/2}(\partial\Omega)$.

The two-dimensional thin domains considered in [20] is given as follows: For $R \geqslant 1$, let Ω_R be a rectangle-shaped domain whose boundary consists of three parts, say

$$\partial\Omega_R = \Gamma_R^+ \cup \Gamma_R^- \cup \Gamma_R^s,$$

where the top and bottom are

$$\Gamma_R^+ = [-R, R] \times \{1\}, \quad \Gamma_R^- = [-R, R] \times \{-1\},$$

and the side Γ_R^s consists of the left and right sides, namely, $\Gamma_R^s = \Gamma_R^l \cup \Gamma_R^r$, where Γ_R^l and Γ_R^r are translates of curves Γ^l and Γ^r connecting points $(0, 1)$ and $(0, -1)$, namely, $\Gamma_R^l = \Gamma^l - (R, 0)$ and $\Gamma_R^r = \Gamma^r + (R, 0)$. If both Γ^l and Γ^r are line segments, Ω_R is a rectangle. The boundary $\partial\Omega_R$ is assumed to be Lipschitz continuous. We say that the domain Ω_R is thin because the scaled domain $R^{-1}\Omega_R$ is thin (like a needle) and its NP spectrum is the same as that of Ω_R (NP spectrum is dilation invariant).

The following theorem is the main result of [19].

Theorem 6.1 *If $\{R_j\}$ be an increasing sequence such that $R_j \to \infty$ as $j \to \infty$, then*

$$\overline{\cup_{j=1}^{\infty} \sigma(\mathcal{K}_{\partial\Omega_{R_j}})} = [-1/2, 1/2]. \tag{44}$$

If Ω_R is a rectangle, then $\sigma_{ess}(\mathcal{K}_{\partial\Omega_R}) = [-1/4, 1/4]$ as we discussed in Sect. 5. Theorem 6.1 says that as R increase to ∞, more and more eigenvalues appear outside

$[-1/4, 1/4]$ and their totality densely fills up intervals $[-1/2, -1/4] \cup [1/4, 1/2]$. This is in accordance with the computational result in [36]. What is surprising is that (44) holds regardless of the choice of the sequence R_j.

Theorem 6.1 is proved as follows. The NP operator on $\partial\Omega_R$ behaves like $1/2$ times the one-dimensional Poisson integral. Since the Fourier transform of the Poisson kernel is $e^{-2\pi t|\xi|}$, the Poisson integral has $[0, 1]$ as its essential spectrum. Using this fact, one can construct a function $\varphi_R \in H^{-1/2}(\partial\Omega_R)$ such that

$$\lim_{R\to\infty} \frac{\|(\lambda I - \mathcal{K}_{\partial\Omega_R})[\varphi_R]\|_*}{\|\varphi_R\|_*} = 0 \tag{45}$$

for each $\lambda \in (0, 1/2]$. It implies that $[0, 1/2] \subset \overline{\bigcup_{j=1}^{\infty}\sigma(\mathcal{K}_{\partial\Omega_{R_j}})}$. In fact, if $\lambda \notin \overline{\bigcup_{j=1}^{\infty}\sigma(\mathcal{K}_{\partial\Omega_{R_j}})}$, there is $\delta > 0$ such that $[\lambda - \delta, \lambda + \delta] \cap (\overline{\bigcup_{j=1}^{\infty}\sigma(\mathcal{K}_{\partial\Omega_{R_j}})}) = \varnothing$. Thus,

$$\|(\lambda I - \mathcal{K}_{\partial\Omega_R})[\varphi]\|_* \lesssim \|\varphi\|_*$$

for all φ, contradicting (45). Since NP spectrum on planar domain is symmetric with respect to 0, we have (44).

The property (44) seems to be a generic property of thin planar domains. For example, thin ellipses enjoy it. In fact, if E_j, $j = 1, 2, \ldots$, is the ellipse defined by $x_1^2/a_j^2 + x_2^2/b_j^2 < 1$ and $\mathcal{K}_{\partial E_j}$ is the corresponding NP operator, where a_j and b_j are positive numbers such that $b_j < a_j$ for all j and $b_j/a_j \to 0$ as $j \to \infty$, then

$$\overline{\bigcup_{j=1}^{\infty}} \sigma(\mathcal{K}_{\partial E_j}) = [-1/2, 1/2]. \tag{46}$$

Since eigenvalues of $\mathcal{K}_{\partial E_j}$ are explicitly given by $\pm 1/2(a_j - b_j)^n/(a_j + b_j)^n$, $n = 1, 2, \ldots$, (see [1]), the proof of (46) is not difficult, and a short proof can be found in [19].

Let us now move to the three-dimensional thin domains. There are two different kinds of thinness in three dimensions: thin and long (like prolate spheroids), thin and flat (like oblate ellipsoids). Their NP spectral structures are different as we see below.

Let Π_R be the prolate spheroid defined by, for $R \geqslant 1$,

$$\Pi_R := \left\{ (x_1, x_2, x_3) : x_1^2 + x_2^2 + \frac{x_3^2}{R^2} < 1 \right\}.$$

Let a_j $(j = 1, 2)$ be positive numbers. Let Ω_R be the oblate ellipsoid defined by

$$\Omega_R := \left\{ (x_1, x_2, x_3) : \frac{x_1^2}{(a_1 R)^2} + \frac{x_2^2}{(a_2 R)^2} + x_3^2 < 1 \right\}.$$

If $a_1 = a_2$, then Ω_R is an oblate spheroid.

The following theorems are obtained in [17].

Theorem 6.2 *If R_j is a sequence of numbers such that $R_j \geqslant 1$ for all j and $R_j \to \infty$ as $j \to \infty$, then*

$$\overline{\bigcup_{j=1}^{\infty} \sigma(\mathcal{K}_{\partial \Pi_{R_j}})} = [0, 1/2]. \tag{47}$$

Theorem 6.3 *If R_j is a sequence of positive numbers such that $R_j \to \infty$ as $j \to \infty$, then*

$$\overline{\bigcup_{j=1}^{\infty} \sigma(\mathcal{K}_{\partial \Omega_{R_j}})} = [-1/2, 1/2]. \tag{48}$$

Theorem 6.2 shows that totality of eigenvalues of $\mathcal{K}_{\partial \Pi_{R_j}}$ is dense in $[0, 1/2]$ regardless of choice of the sequence R_j as long as $1 \leqslant R_j \to \infty$. As mentioned before, it is proved in [3] that there is no negative eigenvalue on prolate spheroids. Theorem 6.3 shows that totality of eigenvalues of $\mathcal{K}_{\partial \Omega_{R_j}}$ is dense in $[-1/2, 1/2]$. This is rather surprising since, as mentioned in Theorem 4.4, $\mathcal{K}_{\partial \Omega_{R_j}}$ admits at most finitely many negative eigenvalues (since Ω_R is strictly convex). However, (48) says that negative eigenvalues in $\bigcup_{j=1}^{\infty} \sigma(\mathcal{K}_{\partial \Omega_{R_j}})$ are dense in $[-1/2, 0]$.

There are many significant works on the NP spectrum on ellipsoids, [2–4, 60, 70] to mention a few. However, it is unlikely that Theorems 6.2 and 6.3 can be proved using those results.

Like two-dimensional case, Theorems 6.2 and 6.3 are proved by investigating the limiting behaviour of the NP operators as $R \to \infty$. It is proved that the NP operator on Π_R converges (on some test functions) to the one-dimensional convolution operator $L * f$ as $R \to \infty$, where

$$L(t) := \frac{1}{2\pi} \int_0^{\pi} \frac{1 - \cos\theta}{[(2 - 2\cos\theta) + t^2]^{3/2}} d\theta.$$

It is then proved that the Fourier transform of L has values in $(0, 1/2]$ and hence the convolution operator has continuous spectrum $[0, 1/2]$. Using this fact, a sequence of functions satisfying (45) is constructed. Then the inclusion

$$[0, 1/2] \subset \overline{\bigcup_{j=1}^{\infty} \sigma(\mathcal{K}_{\partial \Pi_{R_j}})} \tag{49}$$

follows. The opposite inclusion is proved in [3].

The NP operator on oblate ellipsoids has two pieces defined on the upper and lower parts of ellipsoids. It is proved that each piece converges to $1/2$ times the two-dimensional Poisson integral. The Poisson integral operator has continuous spectrum $[0, 1]$. By choosing proper signs on the upper and lower parts, a sequence of functions satisfying (45) is constructed for $\lambda \in [-1/2, 1/2]$ ($\lambda \neq 0$) which yields Theorem 6.3.

A natural question arises: whether Theorem 6.2 holds for cylinder-like convex domains or even prolate ellipsoids. One can show that (49) holds for such domains.

But we do not know if the reverse inclusion is true. We do not know negative NP eigenvalues, if any, on prolate spheroids, disappear eventually if they become thinner.

The property (48) seems to be a generic property of thin, flat domains. To demonstrate it, a typical thin, flat domain is considered. To define such a domain, let U be a bounded planar domain with the Lipschitz continuous boundary ∂U. Let Φ be the domain in $\mathbb{R}^3$ whose boundary consists of three pieces, namely,

$$\partial \Phi = \Sigma^+ \cup \Sigma^- \cup \Sigma^s$$

where the top and bottom are given by $\Sigma^\pm = U \times \{\pm 1\}$ and Σ^s is a surface connecting $\partial U \times \{+1\}$ and $\partial U \times \{-1\}$. We assume that $\partial \Phi$ is Lipschitz continuous. For $R > 0$ let

$$\Phi_R := \{(Rx_1, Rx_2, x_3) : (x_1, x_2, x_3) \in \Phi\}. \tag{50}$$

The following theorem is proved in [17].

Theorem 6.4 *If R_j is a sequence of positive numbers such that $R_j \to \infty$ as $j \to \infty$, then*

$$\overline{\bigcup_{j=1}^{\infty} \sigma(\mathcal{K}_{\partial \Phi_{R_j}})} = [-1/2, 1/2].$$

7 Analysis of Field Concentration

In a composite which consists of inclusions of different material properties and matrix, some inclusions are located close to each other, and a strong stress may occur in the region between closely located inclusions. During last three decades or so, there has been significant progress in quantitative analysis of stress or field concentration about which an extensive survey has been made and several open problems are discussed in [42]. In this section, we briefly review some of them in connection with the NP spectral theory.

Let us mention what the problem is with a brief review of results and how it is related to the spectral theory of the NP operator. Let the domain Ω consist of two closely located but disjoint domains D_1 and D_2, namely, $\Omega = D_1 \cup D_2$. Let ϵ be the distance between D_1 and D_2. Let k_j be the conductivity of D_j for $j = 1, 2$, while that of $\mathbb{R}^d \setminus (D_1 \cup D_2)$ is assumed to be 1. So the conductivity distribution for this section is given by

$$\epsilon = k_1 \chi_{D_1} + k_2 \chi_{D_2} + \chi_{\mathbb{R}^d \setminus (D_1 \cup D_2)}.$$

The conductivities k_1 and k_2 are different from 1 and allowed to be 0 or ∞. The conductivity being ∞ means that the inclusion is perfectly conducting, and 0 means insulating.

We consider the problem (2). The problem is to derive estimates for ∇u (and higher order derivatives) in terms of ϵ (and k_1, k_2, if possible) as ϵ tends to 0. Another problem is to characterize asymptotically the singularity of ∇u. The asymptotic characterization, as ϵ tends to 0, means a decomposition of the form

$$u = s + r, \tag{51}$$

where s is the singular part, namely, ∇s carries the full information of the singularity of ∇u, while r is a regular part, namely, ∇r is bounded.

When D_1 and D_2 are disks, an optimal estimate for the gradient has been derived in [11, 12]. It is extended to higher order derivatives in [31] (see also [39]):

$$\|u\|_{n,U} \lesssim \left(4\lambda_1\lambda_2 - 1 + \sqrt{\epsilon}\right)^{-n} \tag{52}$$

provided that $(k_1 - 1)(k_2 - 1) > 0$. Here U be a bounded set containing $\overline{D_1 \cup D_2}$, $\|u\|_{n,U}$ denotes the piecewise C^n norm on U, namely,

$$\|u\|_{n,U} := \|u\|_{C^n(\overline{D_1})} + \|u\|_{C^n(\overline{D_2})} + \|u\|_{C^n(U \setminus \Omega)},$$

and

$$\lambda_j := \frac{k_j + 1}{2(k_j - 1)}, \quad j = 1, 2. \tag{53}$$

If $k_1 = k_2 = \infty$ (or $k_1 = k_2 = 0$), then the estimate (52) yields

$$|\nabla u(z)| \lesssim \epsilon^{-1/2}. \tag{54}$$

This estimate has been extended to strictly convex inclusions in two dimensions (more precisely, strictly convex near the unique points on ∂D_1 and ∂D_2 of the shortest distance) [74]. In three dimensions, the optimal estimate for ∇u has been obtained in [23]: If $k_1 = k_2 = \infty$ and inclusions are strictly convex inclusions, then

$$|\nabla u(z)| \lesssim \frac{1}{\epsilon |\ln \epsilon|}.$$

However, despite important progress made in [24, 57, 72, 75], the insulating case $(k_1 = k_2 = 0)$ in three dimensions remains unsolved.

The asymptotic characterization of the form (51) for planar strictly convex domains has been obtained in [8, 43, 48]. Moreover, it is proved in [43] that the regular part r converges to the touching case solution as $\epsilon \to 0$ and the singular part disappear as soon as the inclusions are touching, namely, $\epsilon = 0$. The asymptotic characterization for three-dimensional balls has been derived in [49, 56].

If $(k_1 - 1)(k_2 - 1) < 0$ and D_1, D_2 are disks, then the following unexpected estimate is obtained in [39]:

$$\|u\|_{n,\Omega} \lesssim \left(4|\lambda_1\lambda_2| - 1 + \sqrt{\epsilon}\right)^{-n+1}. \tag{55}$$

If $k_1 = 0$ and $k_2 = \infty$ (or the other way around), namely, D_1 is an insulator and D_2 is a perfect conductor, then

$$\lambda_1 = -\frac{1}{2}, \quad \lambda_2 = \frac{1}{2}. \tag{56}$$

Thus (55) yields that

$$\|u\|_{n,\Omega} \lesssim \epsilon^{\frac{-n+1}{2}}.$$

In particular, it implies that ∇u is bounded and

$$\|\nabla^2 u\|_{L^\infty(U)} \lesssim \epsilon^{-1/2}.$$

Boundedness of ∇u has been proved in [58]. This estimate is optimal in the sense that there are harmonic functions h such that the opposite inequality holds.

It is an intriguing problem is to extend the results for circular inclusions to inclusions of general shape (or to prove they do not hold on inclusions of general shape). If D_1 is an insulator and D_2 is a perfect conductor, then the corresponding conductivity problem can be expressed as follows:

$$\begin{cases} \Delta u = 0 & \text{in } \mathbb{R}^d \setminus \overline{\Omega}, \\ \partial_\nu u = 0 & \text{on } \partial D_1, \\ u = c & \text{on } \partial D_2, \\ v(x) - h(x) = O(|x|^{-d+1}) & \text{as } |x| \to \infty, \end{cases}$$

where c is a constant to be determined by the additional condition

$$\int_{\partial D_2} \partial_\nu u|_+ = 0.$$

The results in [39] are obtained using the spectral theory of NP operator on two circles. It was possible since NP eigenvalues and eigenfunctions on two circles can be computed explicitly. It may not be possible to apply the NP spectral theory to inclusions of general shape. However, the NP spectral theory may provide some insight to the problem.

The representation (6) of the solution takes the following form if there are two inclusions, namely, if $\Omega = D_1 \cup D_2$:

$$u(x) = h(x) + \mathcal{S}_{\partial D_1}[\varphi_1](x) + \mathcal{S}_{\partial D_2}[\varphi_2](x), \quad x \in \mathbb{R}^d.$$

The continuity of the flux along ∂D_j takes the following form:

$$k_j \partial_\nu (\mathcal{S}_{\partial D_1}[\varphi_1] + \mathcal{S}_{\partial D_2}[\varphi_2])|_- - \partial_\nu (\mathcal{S}_{\partial D_1}[\varphi_1] + \mathcal{S}_{\partial D_2}[\varphi_2])|_+ = (1 - k_j)\partial_\nu h,$$

which can be written in short as

$$\left(\Lambda - \mathcal{K}^*_{\partial\Omega}\right)[\varphi] = \begin{bmatrix} \partial_\nu h|_{\partial D_1} \\ \partial_\nu h|_{\partial D_2} \end{bmatrix}, \tag{57}$$

where

$$\Lambda := \begin{bmatrix} \lambda_1 I & 0 \\ 0 & \lambda_2 I \end{bmatrix}$$

(λ_j is defined by (53)). Here, I is the identity operator. The NP operator $\mathcal{K}^*_{\partial\Omega}$ on $\partial\Omega = \partial D_1 \cup \partial D_2$ is given by

$$\mathcal{K}^*_{\partial\Omega} \begin{bmatrix} \varphi_1 \\ \varphi_2 \end{bmatrix} = \begin{bmatrix} \mathcal{K}^*_{\partial D_1}[\varphi_1] & \partial_\nu \mathcal{S}_{\partial D_2}[\varphi_2]|_{\partial D_1} \\ \partial_\nu \mathcal{S}_{\partial D_1}[\varphi_1]|_{\partial D_2} & \mathcal{K}^*_{\partial D_2}[\varphi_2] \end{bmatrix}$$

for $\varphi = (\varphi_1, \varphi_2) \in H^{-1/2}(\partial\Omega) = H^{-1/2}(\partial D_1) \times H^{-1/2}(\partial D_2)$.
   If $k_1 = 0$ and $k_2 = \infty$, then

$$\Lambda := \frac{1}{2} \begin{bmatrix} -I & 0 \\ 0 & I \end{bmatrix}$$

by (56). The question is if the integral equation and spectral properties of $\mathcal{K}^*_{\partial\Omega}$ can provide some insight why ∇u is bounded and the second derivative (and higher order derivatives) blows up.

   We may be able to grasp the question better by comparing the insulator-conductor case with the conductor-conductor case ($k_1 = k_2 = \infty$) where the conductivity of both inclusions is ∞. In that case, Λ in (57) is given by

$$\Lambda := \frac{1}{2} \begin{bmatrix} I & 0 \\ 0 & I \end{bmatrix}.$$

Since more and more eigenvalues of $\mathcal{K}^*_{\partial\Omega}$ approach $1/2$ as ϵ (the distance between inclusions) tends to 0 as proved in [26, 27], the solution φ to (57) blows up and so does ∇u as ϵ tends to 0.

   A special spectral feature of circular inclusions which enables us to solve the integral equation (57) is that $\varphi = (\varphi_1, \varphi_2)$ and $\psi = (\psi_1, \psi_2)$ are orthogonal to each other if and only if they are separately orthogonal, that is, φ_j and ψ_j are orthogonal to each other for $j = 1, 2$.

8 Conclusion

We review recent development in spectral geometry and analysis of the NP operator in various topics including plasmon resonance and the NP spectral theory, CALR and analysis of surface localization of plasmon, negative eigenvalues and spectrum on tori, spectrum on polygonal domains, spectral structure of thin domains, and analysis of stress in terms of the NP spectral theory. These topics are complementary to topics covered in another survey paper [22].

During the course of review we discuss some problems to be solved. Among them are

- estimates for the surface localization of the plasmon on planar domains with real analytic boundaries and applications to CALR,
- NP spectrum on tori and possible connection to CALR,
- geometric conditions which guarantee existence of singularly continuous spectrum (an example of a domain with non-empty singularly continuous spectrum is already interesting),
- the question on the appearance of more and more eigenvalues on rectangles as their aspect ratios tend to ∞,
- non-existence (or existence) of an NP eigenvalue on triangles,
- an optimal gradient estimate for the insulating problem,
- derivative estimates for the insulator-conductor problem: general shape.

These problems are all quite interesting and may be quite challenging as well.

The NP operator is also used effectively for solving inverse problems, especially in detection of small inclusions, via the notion of generalized polarization tensors. For example, one can see from (6) and (11) that the solution to the problem (2) admits the dipole expansion

$$u(x) = h(x) - M\nabla h(0) \cdot \nabla \Gamma(x) + O(|x|^{-d}), \quad |x| \to \infty,$$

where M is a $d \times d$ matrix called the polarization matrix. It is a signature of the inclusion Ω and can be used to reconstruct some information of Ω. We refer to [9, 10, 41] for that application and some other applications of the NP operator.

If the polarization tensor of an inclusion is zero, then

$$u(x) = h(x) + O(|x|^{-d}), \quad \text{as } |x| \to \infty,$$

which means that the inclusion is invisible or vaguely visible, and hence hard to be detected. A inclusion whose polarization tensor is made to vanish is called a weakly neutral inclusion (or a polarization tensor vanishing structure). A homogeneous simply connected domain cannot be weakly neutral. But, weakly neutral inclusions may be attained by coating simply connected domains (see [44–46]). We refer to another survey paper [40] for the study on weakly neutral inclusions (as well as neutral inclusions) and related over-determined problem for confocal ellipsoids which is a quite challenging mathematical problem.

References

1. L.V. Ahlfors, Remarks on the Neumann-Poincaré integral equation. Pacific J. Math. **3**, 271–280 (1952)
2. J.F. Ahner, On the eigenvalues of the electrostatic integral operator. II. J. Math. Anal. Appl. **181**, 328–334 (1994)
3. J.F. Ahner, R.F. Arenstorf, On the eigenvalues of the electrostatic integral operator. J. Math. Anal. Appl. **117**, 187–197 (1986)
4. J.F. Ahner, V.V. Dyakin, V.Y. Raevskii, New spectral results for the electrostatic integral operator. J. Math. Anal. Appl. **185**, 391–402 (1994)
5. H. Ammari, G. Ciraolo, H. Kang, H. Lee, G.W. Milton, Spectral theory of a Neumann-Poincaré-type operator and analysis of cloaking due to anomalous localized resonance. Arch. Rational Mech. Anal. **208**, 667–692 (2013)
6. H. Ammari, G. Ciraolo, H. Kang, H. Lee, G.W. Milton, Anomalous localized resonance using a folded geometry in three dimensions. Proc. R. Soc. A **469**, 20130048 (2013)
7. H. Ammari, G. Ciraolo, H. Kang, H. Lee, G.W. Milton, Spectral theory of a Neumann-Poincaré-type operator and analysis of anomalous localized resonance II. Contemporary Math. **615**, 1–14 (2014)
8. H. Ammari, G. Ciraolo, H. Kang, H. Lee, K. Yun, Spectral analysis of the Neumann-Poincaré operator and characterization of the stress concentration in anti-plane elasticity. Arch. Rational. Mech. Anal. **208**, 275–304 (2013)
9. H. Ammari, H. Kang, *Reconstruction of Small Inhomogeneities from Boundary Measurements.* Lecture Notes in Mathematics, vol. 1846. (Springer, Berlin, 2004)
10. H. Ammari, H. Kang, *Polarization and Moment Tensors with Applications to Inverse Problems and Effective Medium Theory*, Applied Mathematical Sciences, vol. 162. (Springer, New York, 2007)
11. H. Ammari, H. Kang, H. Lee, J. Lee, M. Lim, Optimal estimates for the electrical field in two dimensions. J. Math. Pures Appl. **88**, 307–324 (2007)
12. H. Ammari, H. Kang, M. Lim, Gradient estimates for solutions to the conductivity problem. Math. Ann. **332**, 277–286 (2005)
13. H. Ammari, P. Millien, M. Ruiz, H. Zhang, Mathematical analysis of plasmonic nanoparticles: the scalar case. Arch. Rational Mech. Anal. **224**(2), 597–658 (2017)
14. H. Ammari, P. Millien, S. Yu, H. Zhang, Mathematical analysis of plasmonic resonances for nanoparticles: the full Maxwell equations. Jour. Diff. Equ **261**(6), 3615–3669 (2016)
15. K. Ando, Y.-G. Ji, H. Kang, D. Kawagoe, Y. Miyanishi, Spectral structure of the Neumann-Poincaré operator on tori. Ann. I. H. Poincare-AN **36**, 1817–1828 (2019)
16. K. Ando, H. Kang, Analysis of plasmon resonance on smooth domains using spectral properties of the Neumann-Poincaré operator. Jour. Math. Anal. Appl. **435**, 162–178 (2016)
17. K. Ando, H. Kang, S. Lee, Y. Miyanishi, Spectral structure of the Neumann–Poincaré operator on thin ellipsoids and flat domains. arXiv:2110.04716
18. K. Ando, H. Kang, H. Liu, Plasmon resonance with finite frequencies: a validation of the quasi-static approximation for diametrically small inclusions. SIAM J. Appl. Math. **76**, 731–749 (2016)
19. K. Ando, H. Kang, Y. Miyanishi, Exponential decay estimates of the eigenvalues for the Neumann-Poincaré operator on analytic boundaries in two dimensions. J. Integr. Equ. Appl. **30**, 473–489 (2018)
20. K. Ando, H. Kang, Y. Miyanishi, Spectral structure of the Neumann–Poincaré operator on thin domains in two dimensions. J. Anal. Math. arXiv:2006.14377
21. K. Ando, H. Kang, Y. Miyanishi, T. Nakazawa, Surface localization of plasmons in three dimensions and convexity. SIAM J. Appl. Math. **81**(3), 1020–1033 (2021)
22. K. Ando, H. Kang, Y. Miyanishi, M. Putinar, Spectral analysis of Neumann-Poincaré operator. Rev. Roumaine Math. Pures Appl. **ILXV**, 545–575 (2021)

23. E. Bao, Y.Y. Li, B. Yin, Gradient estimates for the perfect conductivity problem. Arch. Rational Mech. Anal. **193**, 195–226 (2009)
24. E. Bao, Y.Y. Li, B. Yin, Gradient estimates for the perfect and insulated conductivity problems with multiple inclusions. Commun. Part. Diff. Eq. **35**, 1982–2006 (2010)
25. J.W. Bates, On toroidal Green's functions. J. Math. Phys. **38**, 3679–3691 (1997)
26. E. Bonnetier, F. Triki, Pointwise bounds on the gradient and the spectrum of the Neumann-Poincaré operator: The case of 2 discs. Contemporary Math. **577**, 79–90 (2012)
27. E. Bonnetier, F. Triki, On the spectrum of Poincaré variational problem for two close-to-touching inclusions in 2D. Arch. Rational Mech. Anal. **209**, 541–567 (2013)
28. D. Chung, H. Kang, K. Kim, H. Lee, Cloaking due to anomalous localized resonance in plasmonic structures of confocal ellipses. SIAM J. Appl. Math. **74**, 1691–1707 (2014)
29. R.R. Coifman, A. McIntosh, Y. Meyer, L'intégrale de Cauchy définit un opérateur borné sur L^2 pour les courbes lipschitziennes. Ann. Math. **116**, 361–387 (1982)
30. G. Dassios, *Ellipsoidal Harmonics* Theory and Applications. (Cambridge University Press, Cambridge, 2012)
31. H. Dong, H. Li, Optimal estimates for the conductivity problem by Green's function method. Arch. Rational Mech. Anal. **231**, 1427–1453 (2019)
32. Yu.V. Egorov, M.A. Shubin (eds.), *Partial Differential Equations VI: Elliptic and Parabolic Operators*, Encyclopaedia of Mathematical Sciences, vol. 63. (Springer, Berlin, 1994)
33. D. Grieser, The plasmonic eigenvalue problem. Rev. Math. Phys. **26**, 1450005 (2014)
34. K.E. Gustafson, D.K.M. Rao, *Numerical Range: The Field of Values of Linear Operators and Matrices* (Springer, New-York, 1997)
35. E. Heine, *Theorie der Kugelfunctionen und der verwandten Functionen, Berlin* (Druck und Verlag von G, Reimer, 1878)
36. J. Helsing, H. Kang, M. Lim, Classification of spectra of the Neumann-Poincaré operator on planar domains with corners by resonance. Ann. I. H. Poincare-AN **34**, 991–1011 (2017)
37. J. Helsing, K.-M. Perfekt, The spectra of harmonic layer potential operators on domains with rotationally symmetric conical points. J. Math. Pures Appl. **118**, 235–287 (2018)
38. Y.-G. Ji, H. Kang, A concavity condition for existence of a negative value in Neumann-Poincaré spectrum in three dimensions. Proc. Amer. Math. Soc. **147**, 3431–3438 (2019)
39. Y.-G. Ji, H. Kang, Spectrum of the Neumann-Poincaré operator and optimal estimates for transmission problems in presence of two circular inclusions. Int. Math. Res. Notices (2022). https://doi.org/10.1093/imrn/rnac057
40. Y. Ji, H. Kang, X. Li, S. Sakaguchi, Neutral inclusions, weakly neutral inclusions, and an over-determined problem for confocal ellipsoids, in *Geometric Properties for Parabolic and Elliptic PDE's*. Springer INdAM series, vol. 47 (2021), pp. 151–181
41. H. Kang, Layer potential approaches to interface problems, in *Inverse Problems and Imaging*, Panoramas et Syntheses 44, Societe Mathematique de France (2014)
42. H. Kang, Quantitative analysis of field concentration in presence of closely located inclusions of high contrast, in *Proceedings ICM 2022*, to appear
43. H. Kang, H. Lee, K. Yun, Optimal estimates and asymptotics for the stress concentration between closely located stiff inclusions. Math. Annalen **363**, 1281–1306 (2015)
44. H. Kang, X. Li, Construction of weakly neutral inclusions of general shape by imperfect interfaces. SIAM J. Appl. Math. **79**, 396–414 (2019)
45. H. Kang, X. Li, S. Sakaguchi, Existence of coated inclusions of general shape weakly neutral to multiple fields in two dimensions. Appl. Anal. (2020). https://doi.org/10.1080/00036811. 2020.1781821
46. H. Kang, X. Li, S. Sakaguchi, Polarization tensor vanishing structure of general shape: existence for small perturbations of balls. Asymptot. Anal. **125**, 101–132 (2021)
47. H. Kang, M. Lim, S. Yu, Spectral resolution of the Neumann-Poincaré operator on intersecting disks and analysis of plasmon resonance. Arch. Rational Mech. Anal. **226**, 83–115 (2017)
48. H. Kang, M. Lim, K. Yun, Asymptotics and computation of the solution to the conductivity equation in the presence of adjacent inclusions with extreme conductivities. J. Math Pures Appl. **99**, 234–249 (2013)

49. H. Kang, M. Lim, K. Yun, Characterization of the electric field concentration between two adjacent spherical perfect conductors. SIAM J. Appl. Math. **74**, 125–146 (2014)
50. H. Kang, J.-K. Seo, Recent progress in the inverse conductivity problem with single measurement, in *Inverse Problems and Related Fields*. (CRC Press, Boca Raton, FL, 2000), pp. 69–80
51. O.D. Kellogg, *Foundations of Potential Theory*, Dover, New York, 1953 (Reprint from the first edition of Die Grundlehren der Mathematischen Wissenschaften), vol. 31. (Springer, Berlin-New York, 1929)
52. D. Khavinson, M. Putinar, H.S. Shapiro, Poincaré's variational problem in potential theory. Arch. Rational Mech. Anal. **185**(1), 143–184 (2007)
53. M. de León-Contreras, K.-M. Perfekt, The quasi-static plasmonic problem for polyhedra. arXiv:2103.13071
54. W. Li, K.-M. Perfekt, S.P. Shipman, Infinitely many embedded eigenvalues for the Neumann-Poincaré operator in 3D. SIAM J. Math. Anal. https://doi.org/10.1137/21M1400365
55. W. Li, S.P. Shipman, Embedded eigenvalues for the Neumann-Poincaré operator. J. Integr. Equ. Appl. **31**(4), 505–534 (2019)
56. H. Li, F. Wang, L. Xu, Characterization of electric fields between two spherical perfect conductors with general radii in 3D. J. Diff. Equ. **267**, 6644–6690 (2019)
57. Y.Y. Li, Z. Yang, Gradient estimates of solutions to the insulated conductivity problem in dimension greater than two, preprint. arXiv:2012.14056
58. M. Lim, S. Yu, Asymptotics of the solution to the conductivity equation in the presence of adjacent circular inclusions with finite conductivities. J. Math. Anal. Appl. **421**, 131–156 (2015)
59. F.C. Marques, A. Neves, The Willmore conjecture. arXiv: 1409.7664
60. E. Martensen, A spectral property of the electrostatic integral operator. J. Math. Anal. Appl. **238**, 551–557 (1999)
61. R.C. McPhedran, G.W. Milton, A review of anomalous resonance, its associated cloaking, and superlensing. C. R. Phys. **21**, 409–423 (2020)
62. G.W. Milton, N.-A.P. Nicorovici, On the cloaking effects associated with anomalous localized resonance. Proc. R. Soc. A **462**, 3027–3059 (2006)
63. Y. Miyanishi, Weyl's law for the eigenvalues of the Neumann–Poincaré operators in three dimensions: Willmore energy and surface geometry. arXiv:1806.03657
64. Y. Miyanishi, G. Rozenblum, Eigenvalues of the Neumann-Poincaré operator in dimension 3: Weyl's law and geometry. Algebra i Analiz **31**(2), 248–268 (2019); reprinted in St. Petersburg Math. J. **31**(2), 371–386 (2020)
65. N.-A.P. Nicorovici, R.C. McPhedran, G.W. Milton, Optical and dielectric properties of partially resonant composites. Phys. Rev. B **49**, 8479–8482 (1994)
66. K.-M. Perfekt, Plasmonic eigenvalue problem for corners: limiting absorption principle and absolute continuity in the essential spectrum. J. Math. Pures Appl. **145**(9), 130–162 (2021)
67. K.-M. Perfekt, M. Putinar, Spectral bounds for the Neumann-Poincaré operator on planar domains with corners. J. d'Analyse Math. **124**, 39–57 (2014)
68. K.-M. Perfekt, M. Putinar, The essential spectrum of the Neumann-Poincaré operator on a domain with corners. Arch. Rational Mech. Anal. **223**, 1019–1033 (2017)
69. M. Reed, B. Simon, *Methods of Modern Mathematical Physics I: Functional Analysis* (Academic Press, 1972)
70. S. Ritter, The spectrum of the electrostatic integral operator for an ellipsoid, in *Inverse Scattering and Potential Problems in Mathematical Physics*, ed. by R.F. Kleinman, R. Kress, E. Marstensen (Lang, Frankfurt/Bern, 1995), pp. 157–167
71. H.S. Shapiro, *The Schwarz Function and its Generalization to Higher Dimensions* University of Arkansas Lecture Notes in the Mathematical Sciences 9. (A Wiley-Interscience Publication, Wiley, New York, 1992)
72. B. Weinkove, The insulated conductivity problem, effective gradient estimates and the maximum principle. Math. Ann. **421** (2022). https://doi.org/10.1007/S00208-201-02314-3
73. K. Yosida, *Functional Analysis*, 4th edn. (Springer, Berlin, 1974)

74. K. Yun, Estimates for electric fields blown up between closely adjacent conductors with arbitrary shape. SIAM J. Appl. Math. **67**, 714–730 (2007)
75. K. Yun, An optimal estimate for electric fields on the shortest line segment between two spherical insulators in three dimensions. J. Diff. Equ. **261**, 148–188 (2016)

Sausages and Butcher Paper

Danny Calegari

Abstract For each $d > 1$ the *shift locus of degree d*, denoted $\mathcal{S}_d$, is the space of normalized degree d polynomials in one complex variable for which every critical point is in the attracting basin of infinity under iteration. It is a complex analytic manifold of complex dimension $d - 1$. We are able to give an explicit description of $\mathcal{S}_d$ as a contractible complex of spaces, and to describe the pieces in two quite different ways: (1) (combinatorial): in terms of dynamical extended laminations; or (2) (algebraic): in terms of certain explicit 'discriminant-like' affine algebraic varieties. From this structure one may deduce numerous facts, including that $\mathcal{S}_d$ has the homotopy type of a CW complex of real dimension $d - 1$; and that $\mathcal{S}_3$ and $\mathcal{S}_4$ are $K(\pi, 1)$s. The method of proof is rather interesting in its own right. In fact, along the way we discover a new class of complex surfaces (they are complements of certain singular curves in $\mathbb{C}^2$) which are homotopic to locally CAT(0) complexes; in particular they are $K(\pi, 1)$s.

1 Introduction

For each $d > 1$ the *shift locus of degree d*, denoted $\mathcal{S}_d$, is the space of normalized (i.e. monic with roots summing to zero) degree d polynomials in one complex variable for which every critical point is in the attracting basin of infinity under iteration. A polynomial in $\mathcal{S}_d$ is called a *shift polynomial*. These are the polynomials whose dynamics are the easiest to understand; perhaps in compensation, their parameter spaces appear to be extremely complicated. Much is known about the geometry and topology of $\mathcal{S}_d$ and much is still mysterious.

The main point of this paper is to describe a canonical decomposition of $\mathcal{S}_d$ (and some equivalent spaces) into pieces, giving $\mathcal{S}_d$ the explicit structure of a 'complex of spaces' over a rather nice space (a contractible complex which is a sort of incomplete $\tilde{A}_{d-2}$ building) and to give two, quite different, descriptions of the pieces.

D. Calegari (✉)
Department of Mathematics, University of Chicago, Chicago, IL 60637, USA
e-mail: dannyc@math.uchicago.edu

One description is combinatorial, in terms of certain iterated fiber bundles resp. their orbifolded quotients that we call *monkey prisms* resp. *monkey turnovers*. In this description, the fibers and their monodromy are encoded quite explicitly in objects called *dynamical elaminations*; the word 'elamination' here is shorthand for 'extended lamination'—a lamination with 'extra' structure. Elaminations are related to the sorts of laminations used elsewhere in holomorphic dynamics (see e.g. [28]) but are in some ways quite different. Their definitions and basic properties are given in Sect. 3, and they are a key tool throughout the remainder of the paper.

The other description is algebraic, in terms of certain complex affine varieties, which arise as moduli spaces of maps between infinite nodal genus 0 surfaces called *sausages*. The relationship between sausages and shift polynomials is of an essentially topological nature, so that although both objects and their moduli spaces carry natural complex analytic structures, the maps between them do *not* respect this structure. This seems to be unavoidable: the shift locus is a highly transcendental object, whereas the moduli spaces we construct are algebraic.

One interesting consequence of this relationship between these two ways of seeing the shift locus is that information which is obscure on one side can become transparent on the other. Here is an example. In degree 3, the Shift Locus can be described (up to homotopy) as a space obtained from the 3-sphere by drilling out a trefoil knot, and gluing in a bundle over S^1 whose fiber is a disk minus a Cantor set. This Cantor set can be thought of as an infinite nested intersection $K = \cap E_n$ of subsets of the disk, where each E_n is itself a finite union of disks. The monodromy permutes each E_n, and it is a fact (Theorem 7.9) that the orbits of this permutation are cycles whose lengths are powers of 2. The only proof of this that I know is to interpret the permutation action of the monodromy as an action on the roots of a certain polynomial obtained by iterated quadratic extensions.

The value of mathematical machinery is that it can prove theorems whose statement does not mention the machinery. As a consequence of our structure theorems we are able to deduce some facts about the topology of the shift locus, especially in low degrees. In particular:

Homotopy Theorem S_d *has the homotopy type of a* $(d - 1)$-*complex (i.e. a complex of half the real dimension of* S_d *as a manifold). For* $d = 3$ *or* 4 *it is a* $K(\pi, 1)$. *For* $d = 3$ *it is homotopic to a CAT(0) 2-complex.*

This is an amalgamation of Theorems 7.5, 8.7 and 8.12. In fact, it is plausible that S_d is a $K(\pi, 1)$ in every degree.

In degree 3 we are able to give an extremely explicit description. S_3 is homeomorphic to a product $X_3 \times \mathbb{R}$ where X_3 is a 3-manifold obtained from S^3 by drilling out a right-handed trefoil, and gluing in a bundle $D_\infty \to N_\infty \to S^1$ where each fiber $D_\infty(t)$ over $t \in S^1$ is a disk minus a Cantor set. In fact, we are able to give a completely explicit description of D_∞ and its monodromy in terms of an object called the *Tautological Elamination*. There is one Tautological Elamination $\Lambda_T(t)$ for each t. These elaminations vary continuously in the so-called collision topology (defined in Sect. 3.2), and the $D_\infty(t)$ are obtained by an operation called *pinching*. Finally, the monodromy is completely described by the formula $\mathcal{F}_t \Lambda_T(s) = \Lambda_T(s+t)$ where $\mathcal{F}_t$ is an explicit flow on the space of elaminations.

In words: the monodromy on D_∞ is the composition of infinitely many fractional Dehn twists in a disjoint collection of circles, associated to the elamination Λ_T in a concrete manner. The combinatorics of Λ_T is rather complicated and beautiful; Theorem 7.9 and Sect. 9.5 describe some of its properties.

One intermediate result that we believe is interesting in its own right, is the discovery of a new class of affine complex surfaces which are $K(\pi, 1)$s:

Regular Value Theorem *Let Y_n be the space of degree 3 polynomials $z^3 + pz + q$ for which n specific complex values (e.g. the nth roots of unity) are regular values. Then Y_n is homotopic to a locally CAT(0) complex, and consequently is a $K(\pi, 1)$.*

Even the case $n = 2$ is new, so far as we know.

1.1 Apology

'Butcher' in the title of this paper and throughout is a rather inelegant pun on the name Böttcher which, Curt McMullen informs me, translates to *cooper* in English (i.e. a maker of casks). However etymologically misguided, I have decided to keep 'butcher' for the sake of the sausages.

1.2 Other Work

I would like to compare and connect the constructions and techniques in this paper to prior and ongoing work of other mathematicians. First and foremost I would like to emphasize the resemblance of elements of the theory of dynamical elaminations to the DeMarco–Pilgrim theory of *pictographs* as explained in [22] (to the degree that I understand them). In fact, DeMarco, sometimes in collaboration with Pilgrim or McMullen, has developed a sophisticated and intricate picture of the shift locus over many years and papers; e.g. [20, 21]. The fact that $\mathcal{S}_d$ has the homotopy type of a $(d-1)$-complex follows from DeMarco's thesis [19], where it is proved that $\mathcal{S}_d$ is a Stein manifold. I wish I better understood the relationship between her work and the point of view we develop here.

Recently, Blokh et al. [2] have developed a theory of laminations to parameterize the pinching of components of (higher degree) Mandelbrot sets. I believe there is a family resemblance of their laminations to the tautological elamination we introduce in Sect. 7.1 and its variants and completions in higher degree, but the precise relationship is unclear.

The significance of configuration-space techniques (e.g. braiding of roots, attractors, etc.) to complex dynamics has been apparent at least since the work of McMullen [25] and Goldberg–Keen [23]. This is a vast story that I only touch on briefly in Sect. 10.

Branner–Hubbard [8], in a tour de force, found a detailed description of much of the parameter space of degree 3 polynomials. In particular, they showed that $\mathcal{S}_3$ (away from a piece with easily understood topology) has the structure of a bundle over a circle (up to homotopy) whose fiber has free fundamental group. This is perfectly parallel to our Theorem 7.4. However, in their theory (which is more concretely tied to polynomials) the monodromy is completely opaque, and the culmination of their description (in Sect. 11.4) is only meant to indicate how formidable an explicit computation would be. Whereas in our theory, we have a completely explicit description of the fiber (it is the disk obtained by pinching the tautological elamination) and the monodromy (rotation by $\mathcal{F}_t$).

2   The Shift Locus

Fix an integer $d > 1$, and let $f(z) = \sum b_j z^{d-j}$ be a complex polynomial of degree d, so that $b_0 \neq 0$. A change of variables $z \to \alpha z + \beta$ with $\alpha \in \mathbb{C}^*$ conjugates f to a polynomial

$$f(z) = \sum \frac{b_j}{\alpha} (\alpha z + \beta)^{d-j} - \frac{\beta}{\alpha} = \alpha^{d-1} b_0 z^d + \alpha^{d-2} (d\beta b_0 + b_1) z^{d-1} + \cdots$$

Setting $\alpha = b_0^{1/(d-1)}$ and $\beta = -b_1/db_0$ we can put f in *normal form*

$$f(z) = z^d + a_2 z^{d-2} + a_3 z^{d-3} + \cdots + a_d$$

There is non-uniqueness in the choice of α; different choices differ by multiplication by a $(d-1)$st root of unity ζ, which multiplies the coefficient a_j by ζ^{d-j-1}.

Definition 2.1 (*Shift locus*) The *shift locus* of degree $d \geq 2$, denoted $\mathcal{S}_d$, is the space of normalized degree d polynomials f for which every critical point of f is in the attracting basin of infinity.

The critical points of f are the roots of f'. To say a point c is in the attracting basin of infinity means that the iterates $c, f(c), f^2(c), \cdots$ converge to infinity.

Note that the property of being in the shift locus is expressed in purely dynamical terms. Thus we could define $\mathcal{S}_d$ to be the space of *conjugacy classes* of polynomials with a certain dynamical property. The relationship between that definition and the one we adopt comes down to an ambiguity of $\mathbb{Z}/(d-1)\mathbb{Z}$ in the representation of a conjugacy class by a normalized polynomial.

The coefficients of a normalized degree d polynomial embed $\mathcal{S}_d$ as a subset of $\mathbb{C}^{d-1}$. It is clear that $\mathcal{S}_d$ is open, since for any polynomial f the punctured disk

$E(R) := \{z : |z| > R\}$ is in the attracting basin of infinity for sufficiently big R (depending continuously on f), and f is in S_d if and only if there is some integer n so that $f^n(c) \in E(R)$ for all critical points c.

Recall the following definition:

Definition 2.2 (*Julia Set*) The *Julia set* J_f of a polynomial f is the closure of the set of repelling periodic orbits of f.

The complement of J_f in the Riemann sphere is the *Fatou* set Ω_f; it is the maximal (necessarily open) set on which f and all its iterates together form a normal family. Actually, it is perhaps more natural to take this to be the definition of the Fatou set, and to define the Julia set to be its complement. The Julia set and the Fatou set are both totally invariant (i.e. $f(J_f) = J_f = f^{-1}(J_f)$ and similarly for Ω_f). The Julia set is always nonempty and perfect. See e.g. Milnor [26], Sect. 4.

Proposition 2.3 *A polynomial f is in the shift locus if and only if the Julia set J_f is a Cantor set on which the action of f is uniformly expanding.*

Proof If J_f is a Cantor set, its complement is connected and is therefore equal to the attracting basin of infinity. If f is uniformly expanding on J_f then $|f'|$ is bounded below on J_f by a positive constant, so J_f can't contain any critical points and f is in the shift locus.

Conversely, suppose f is in the shift locus. Since ∞ is an attracting fixed point, there is a connected neighborhood U of ∞ with $f(U) \subset U$. Because f is a polynomial, ∞ is its own unique preimage under f; it follows by induction that for each n, the set $V_n := f^{-n}(U)$ is both forward-invariant and connected (because each component contains ∞). Because f is in the shift locus, there is an n so that all the critical points are contained in V_n. Let K be the complement of V_n, so that K is a finite union of disks.

Because all the critical points are in V_n, each point in K has exactly d distinct preimages; these vary continuously as a function of K, and since each component D of K is simply-connected, $f^{-1}|D$ has d well-defined continuous branches with disjoint image. By the Schwarz Lemma the branches of f^{-1} are uniformly contracting in the hyperbolic metric on each component of K; thus the diameters of the components of $f^{-n}(K)$ converge (at a geometric rate) to zero, so that $\Lambda := \cap_n f^{-n}(K)$ is totally disconnected and f is uniformly expanding on Λ.

Evidently the complement of Λ is the basin of infinity, so $J_f = \Lambda$. Since J_f is always perfect, it is a Cantor set, and f is uniformly expanding on J_f, as claimed. $\square$

Example 2.4 (*Mandelbrot set*) A quadratic polynomial $z \to z^2 + c$ has 0 as its unique critical point. The set of $c \in \mathbb{C}$ for which 0 is *not* in the basin of infinity of $z \to z^2 + c$ is called the *Mandelbrot set* $\mathcal{M}$; see Fig. 1. Thus $\mathcal{M}$ is the complement of S_2 in $\mathbb{C}$. The connectivity of the Mandelbrot set (proved by Douady and Hubbard [18]) is equivalent to the fact that S_2 is homeomorphic to an (open) annulus.

Fig. 1 The Mandelbrot set
$\mathcal{M}$ (interior in white) is the
complement of $\mathcal{S}_2$ in $\mathbb{C}$

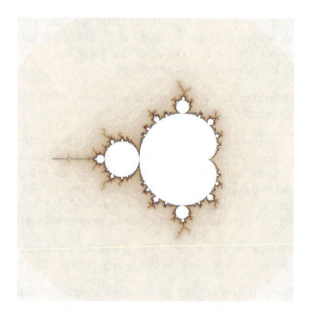

Example 2.5 (*Discriminant complement*) Let $f(z)$ be any degree d polynomial
with distinct roots (i.e. for which 0 is not a critical value). Then $g(z) := \lambda f(z)$ is
(conjugate to a polynomial) in the shift locus for $|\lambda| \gg 1$. To see this, let U be any
neighborhood of infinity for which $f(U)$ does not contain 0. Then for sufficiently
large $|\lambda|$ we have $g(U) \subset U$ (so that U is contained in the attracting basin of infinity
for g). Furthermore, g and f have the same critical points, so for sufficiently large
$|\lambda|$ we have $g(c) \in U$ for every critical point c of g.

We can think of this as showing that near infinity, $\mathcal{S}_d$ is 'nearly equal' to the
complement of the discriminant locus $\mathbb{C}^{d-1} - \Delta$. We shall elaborate on this remark
in the sequel.

Example 2.6 (*Cantor set J_f*) In degree two, J_f is a Cantor set precisely when f is
in the shift locus, but for degree bigger than two it is possible for J_f to be a Cantor
set for f not in the shift locus.

For example, consider the polynomial $f(z) := \alpha z(z - 1)^2$ with α real and positive. The fixed points are 0 and $\beta^{\pm} := 1 \pm \sqrt{1 - (\alpha - 1)/\alpha}$ and the critical points
are $1/3$ and 1. Since $f(1) = 0$, the polynomial f is never in the shift locus. If
$f(1/3) > \beta^+$ then $f^{-1}([0, \beta^+])$ is real and properly contained in $[0, \beta^+]$, and
$J_f = \cap_n f^{-n}([0, \beta^+])$ is a totally real Cantor set. This happens for $\alpha > 9$.

In the limiting case $\alpha = 9$, the Julia set J_f is the real interval $[0, 4/3]$.

Suppose f is in the shift locus, so that J_f is a Cantor set, equal to the complement
of the basin of infinity. Then f has d distinct fixed points, all in J_f.

Because the dynamics of f on J_f is expanding, it is structurally stable there. So if
f_t is a family of polynomials in the shift locus with Julia sets J_{f_t}, there are open sets

$U(t)$ containing J_{f_t} and maps $\varphi_t : U(0) \to U(t)$ conjugating $f_t|U(t)$ to $f_0|U(0)$. In particular, we obtain a *monodromy representation* ρ from the fundamental group $\pi_1(\mathcal{S}_d)$ to the *mapping class group* of $\mathbb{C} -$ Cantor set. This is an example of a so-called *big mapping class group*; see e.g. [30] for background and an introduction to the theory of such groups.

The dynamics of any f on J_f is conjugate to the action of the shift on the space of one-sided sequences in a d letter alphabet; this justifies the name. One way to see this is to take a compact K containing J_f in its interior for which $f|K$ has d inverse branches $f_1, \cdots, f_d$, and the $f_j(K)$ are disjoint subsets of the interior of K. Then J_f is in bijection with the set of right infinite words in the $\{f_j\}$.

The geometry of $\mathcal{S}_d$ is very complicated. For $d = 2$ the space $\mathcal{S}_2$ is the complement in $\mathbb{C}$ of the Mandelbrot set; showing that $\mathcal{S}_2$ is conformal to a punctured disk is equivalent to showing that the Mandelbrot set is connected. The main goal of this paper is to develop tools to describe the topology of $\mathcal{S}_d$ for higher d.

3 Elaminations

In this section we introduce the concept of an *elamination*. Laminations, as introduced by Thurston, are a key tool in low-dimensional geometry, topology and dynamics; see e.g. [27], Sect. 8.5. The reader already familiar with laminations can think of the term 'elamination' as an abbreviation for 'extended lamination', or 'enhanced lamination'—an ordinary lamination with some extra structure.

Elaminations are an essential combinatorial tool that will be used throughout the sequel, especially beginning with Sect. 4, so throughout this section we just spell out the basic theory, deferring the connection to dynamics until the sequel. There are some points of contact between elaminations—and in particular the 'collision topology' on the space $\mathcal{EL}$—to the theory partially developed by Thurston in [29]; but there are many points of difference, and it seems pointless to try to force the two theories into a common framework.

Elaminations (and laminations for that matter) have several more-or-less equivalent identities, and it is useful to be able to move back and forth between them. By abuse of notation, we will often use the same symbol or term to refer to the underlying abstract object or any of its equivalent manifestations.

We fix the following notation here and throughout the rest of the paper: let $\mathbb{D}$ denote the *closed* unit disk in the complex plane $\mathbb{C}$, and let $\mathbb{E} := \mathbb{C} - \mathbb{D}$ denote its *open* exterior.

Definition 3.1 (*Circle Lamination*) A *leaf* is a finite subset of the unit circle of cardinality at least 2. A leaf is *simple* if it consists of 2 points; a leaf of *multiplicity* n consists of $n + 1$ points.

A *circle lamination* is a set of leaves, no two of which have 2 element subsets that are linked. A circle lamination is simple if all its leaves are simple.

Most authors require laminations to be closed in the space of finite subsets of S^1 (in the Hausdorff topology), but we explicitly do *not* require this.

Definition 3.2 (*Geodesic Lamination*) A *simple geodesic leaf* is a complete geodesic in $\mathbb{D}$ with its hyperbolic metric. A *geodesic leaf of multiplicity* $n > 1$ is an ideal $(n + 1)$-gon.

A *geodesic lamination* is a set of geodesic leaves no two of which cross in $\mathbb{D}$. A geodesic lamination is simple if all its leaves are simple.

Every ideal $(n + 1)$-gon in $\mathbb{D}$ determines an unordered set of $n + 1$ endpoints in S^1 and conversely. Two $(n + 1)$-gons in $\mathbb{D}$ cross if and only if two pairs of their endpoints link in S^1. Thus there is a natural correspondence between circle laminations and geodesic laminations.

Definition 3.3 (*Elamination*) For each $z \in \mathbb{E}$ we let $\ell(z)$ denote the straight line segment from $z/|z|$ to z. We call $\ell(z)$ a *radial segment*. The *height* of the segment $\ell(z)$ is $\log(|z|)$.

An *extended leaf* of *height* $h > 0$ is the union of a geodesic leaf in $\mathbb{D}$ (the *vein*) with radial segments in $\mathbb{E}$ (the *tips*) all of height h, attached at the endpoints of the vein. An extended leaf is *simple* if the vein is simple.

An *extended lamination*, or *elamination* for short, is a set of extended leaves with the following properties:

(1) lamination: distinct leaves have distinct veins, and the set of all veins of all leaves forms a geodesic lamination (called the *vein* of the elamination);
(2) properness: there are only finitely many extended leaves with height $\geq \epsilon$ for any $\epsilon > 0$ (thus every elamination has only countably many leaves); and
(3) saturation: to be defined below.

Let us now explain the meaning of saturation. Let Λ be an elamination, and let ℓ be a leaf with height h. Let pq be an oriented edge of ℓ, and let L be the finite set of leaves of Λ on the positive side of pq with height $\geq h$. Let L_p (resp. L_q) denote the subset of L of leaves with an endpoint with the same argument as p (resp. q). Since leaves of Λ do not cross, and distinct leaves have distinct veins, the leaves L_p are ordered by how they separate each other from pq; thus if L_p is nonempty there is a *closest* $\ell_p \in L_p$ to pq (and similarly for L_q). A leaf ℓ_p (resp. ℓ_q) if it exists, is called an *elder sibling* for ℓ at p (resp. at q).

Saturation means the following two conditions hold for every ℓ:

(1) an elder sibling of ℓ has height h' strictly bigger than h; and
(2) if L_p is nonempty so is L_q and vice versa; and furthermore $\ell_p = \ell_q$.

We say that a leaf ℓ is *saturated* by an elder sibling. Another way to say this is that if the vein of ℓ shares one endpoint with the vein of a taller leaf ℓ', and there are no other ℓ'' (also taller than ℓ) in the way, then the vein of ℓ actually shares two endpoints with ℓ'.

3.1 Pinching

Let Λ be an elamination. We define an operation called *pinching* that associates to Λ a Riemann surface Ω obtained from $\mathbb{E}$ by suitable cut and paste along the tips of Λ.

Construction 3.4 (*Pinching*) Let Λ be an elamination. For each leaf λ with multiplicity n and with tips $\sigma_0, \cdots, \sigma_n$ enumerated in cyclic order in S^1, cut open $\mathbb{E}$ along the σ_j and glue the right side of each σ_j to the left side of σ_{j-1} (indices taken mod $n + 1$) by a Euclidean isometry.

The resulting Riemann surface Ω is said to be obtained from Λ by *pinching*. We also write $\Omega = \mathbb{E} \mod \Lambda$.

Lemma 3.5 (Planar) Ω *obtained from an elamination Λ by pinching is planar.*

Proof This is equivalent to the fact that the leaves do not cross. $\square$

By construction, the function $\log | \cdot | : \mathbb{E} \to (0, \infty)$ is preserved under pinching, and therefore descends to a well-defined proper function on Ω that we refer to as the *height function* or sometimes as the *Green's function*, and denote h. Furthermore, $d \arg$ is a well-defined 1-form on Ω, so the level sets of the height function are finite unions of metric graphs. We sometimes denote $d \arg$ by $d\theta$. In fact, the combination $dh + id\theta$ is just the image of $d \log(z)$ on $\mathbb{E}$, which makes sense because this 1-form is preserved by cut-and-paste. By abuse of notation therefore we sometimes write $dh + id\theta = d \log(z)$. This 1-form has a zero of multiplicity m for each leaf of multiplicity m.

Definition 3.6 (*Monkey pants*) A *monkey pants* is a (closed) disk with at least two (open) subdisks removed. If P is a monkey pants, a function $\pi : P \to [t_1, t_2]$ is *monkey Morse* if it is a submersion away from finitely many points in the interior which are all saddles or monkey saddles, and if $\pi^{-1}(t_2)$ is equal to a distinguished boundary component $\partial^+ P$ (the *waist*) and $\pi^{-1}(t_1)$ is equal to the other components $\partial^- P$ (the *cuffs*).

Let Ω be the Riemann surface associated to an elamination. If $0 < t_1 < t_2$ are numbers not equal to the height of any leaf, then $\Omega([t_1, t_2]) := h^{-1}[t_1, t_2] \subset \Omega$ is a monkey pants, and h restricted to $\Omega([t_1, t_2])$ is monkey Morse. There is one saddle point for each simple leaf with height in $[t_1, t_2]$, and one monkey saddle with multiplicity equal to the multiplicity of a non-simple leaf.

Suppose Λ is a finite elamination, which pinches $\mathbb{E}$ to Ω. Then Ω is a plane minus $n + 1$ disks, where n is the number of leaves of Λ counted with multiplicity. If t is the least height of leaves of Λ, then $\Omega((0, t))$ is a disjoint union of $n + 1$ annuli whose inner 'boundary components' (where $h \to 0$) can be compactified by $n + 1$ circles. We refer to this collection of circles as $S^1 \mod \Lambda$. Thus: just as $\mathbb{E}$ is compactified (away from ∞) by S^1, the surface $\mathbb{E} \mod \Lambda$ is compactified (away from ∞) by $S^1 \mod \Lambda$.

3.2 Push over and Amalgamation

Denote the set of elaminations by $\mathcal{EL}$. We would like to define a natural topology on $\mathcal{EL}$. In a nutshell, a family of elaminations Λ_t in $\mathcal{EL}$ varies continuously if and only if the Riemann surfaces $\Omega_t = \mathbb{E} \mod \Lambda_t$ do.

Because of properness, an elamination Λ has only finitely many leaves of height bigger than any positive ϵ. When these leaves have disjoint veins, it is obvious what it means to say that they vary continuously in a family: it just means that the heights and arguments vary continuously.

When two leaves of different heights collide, the shorter leaf becomes *saturated* by the taller (which becomes at that moment its elder sibling); if we continue the motion in the obvious way, the shorter leaf becomes unsaturated as it moves away from the taller leaf, and the net result is that the shorter leaf has been *pushed over* the taller one. The meaning of this is illustrated in Fig. 2.

When two leaves of the same height collide, saturation dictates that they must become amalgamated into a common leaf; see Fig. 3.

We now define a topology on $\mathcal{EL}$ called the *collision topology*.

Definition 3.7 (*Collision Topology*) A family of elaminations Λ_t varies *continuously* in $\mathcal{EL}$ in the collision topology if every finite subset of leaves varies continuously when they are disjoint, and varies by push over or amalgamation when they collide.

The whole point of the collision topology is that it is compatible with pinching.

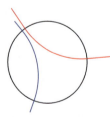  

Fig. 2 Pushing a shorter leaf over a taller one; at the intermediate step the shorter leaf is saturated by the taller one

Fig. 3 When two simple leaves of the same height collide, they amalgamate to form a leaf of multiplicity 2

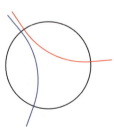 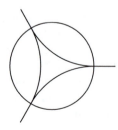

Lemma 3.8 (Continuous quotient) *If Λ_t varies continuously in $\mathcal{EL}$ then Ω_t vary continuously as Riemann surfaces.*

Proof The only thing to check is that push over and amalgamation are continuous under pinching; but this is essentially by definition. ☐

4 Butcher Paper

4.1 Böttcher Coordinates

Let $f(z) := z^d + a_2 z^{d-2} + \cdots + a_d$ be a degree d polynomial in normal form. Lucjan Böttcher, a Polish mathematician who worked in Lvov in the beginning of the 20th century, showed [3] that f is conjugate to $z \to z^d$ in a neighborhood of infinity:

Proposition 4.1 (Böttcher Coordinates) *Let $f(z) := z^d + a_2 z^{d-2} + \cdots + a_d$ be a degree d polynomial in normal form. Then f is holomorphically conjugate to $z \to z^d$ on some neighborhood of infinity.*

For a proof see e.g. Milnor [26], Theorem 9.1.

4.2 Holomorphic 1-Form

Let's let ϕ be the holomorphic conjugacy promised by Proposition 4.1 normalized so that $\phi f \phi^{-1}(z) = z^d$ near infinity. The map ϕ is only defined in a neighborhood of infinity, but we can extend it inductively over larger and larger domains by using the functional equation. Recall that $\mathbb{E}$ denotes the exterior of the closed unit disk in $\mathbb{C}$; i.e. $\mathbb{E}$ is the basin of infinity of $z \to z^d$. The function $\log z$ is not single-valued on $\mathbb{E}$, but its differential dz/z is. The map $z \to z^d$ pulls back dz/z to $d \cdot dz/z$ (we use the notation $d\cdot$ to indicate multiplication by the degree d to distinguish it from the exterior derivative of forms). If we define $\alpha := \phi^* dz/z$ in a neighborhood of infinity, we can extend α uniquely to all of the Fatou set Ω_f by iteratively solving $f^* \alpha = d \cdot \alpha$. Thus α is a holomorphic 1-form on Ω_f with zeroes at the critical points of f and their preimages.

4.3 Horizontal/Vertical Foliations

The real and imaginary parts of α and dz/z give rise to foliations on Ω_f and on $\mathbb{E}$ related by ϕ near infinity. We call these the *horizontal* and the *vertical* foliations respectively.

On $\mathbb{E}$ these foliations are nonsingular; the horizontal leaves are the circles $|z| = $ constant and the vertical leaves are the rays $\arg(z) = $ constant. The corresponding foliations on Ω_f have saddle singularities at simple critical points and their preimages, and monkey saddle singularities at critical points (and their preimages) of multiplicity bigger than one (as roots of f'). Evidently ϕ may be extended by analytic continuation along every nonsingular vertical leaf, and along every singular leaf from infinity until the first singularity. These singularities are critical points and their preimages; this is a proper subset of Ω_f.

4.4 Construction of the Dynamical Elamination

Let $L_f \subset \Omega_f$ be the complement of this (maximal) domain of definition of ϕ, and $L \subset \mathbb{E}$ the complement of $\phi(\Omega_f - L_f)$. These subsets are both closed and backwards invariant. The complements $\Omega_f - L_f$ and $\mathbb{E} - L$ are open, simply connected, and dense. The set L consists of a countable collection of radial segments; in the generic case there are exactly two such segments $\ell(q^{\pm})$ for each critical or pre-critical point p. One may think of $q^{\pm}$ as the 'image' of p under ϕ. If c is a simple critical point with image $v = f(c)$ then $\phi(v)$ will have d preimages under $z \to z^d$, whereas v will only have $d - 1$ preimages under f; the two of the preimages of $\phi(v)$ that correspond to c are $q^{\pm}$.

Example 4.2 If f has real coefficients, ϕ preserves the real axis. Thus the vertical leaves with $\arg(\phi(z)) \in \pi d^{-n}\mathbb{Z}$ consist of the z with $f^n(z)$ real. The polynomial $f(z) := z^3 + 3z + 3^{-1/2}$ has critical points at $\pm i$ with initial forward orbit

$$\pm i \to 3^{-1/2} \pm 2i \to -23 \cdot 3^{-3/2} \approx -4.42635$$

Figure 4 shows some vertical leaves in Ω_f and in $\mathbb{E}$ in the preimage of the negative real axis. L_f and L are in red. The set $L_f \cup J_f$ is a dendrite.

Note that $\arg(\phi(f^2(i))) = \pi$ and $\arg(\phi(f(i))) = \pi/3$. The absolute value $|\phi(i)|$ is well-defined, and equal to approximately 1.18, but $\arg(\phi(i))$ is multi-valued, and takes values $7\pi/9$ and $\pi/9$.

One may repair this multi-valuedness of ϕ by doing cut-and-paste on $\mathbb{E}$: cut open $\mathbb{E}$ along the segments L and reglue edges in pairs, so that each copy of $\ell(q^+)$ is glued to a copy of $\ell(q^-)$ in the unique manner which is orientation-reversing and compatible with the dynamics $z \to z^d$. The result is a new Riemann surface Ω on which the map $z \to z^d$ on $\mathbb{E} - L$ extends uniquely to a holomorphic degree d map $F : \Omega \to \Omega$ and for which $\phi : \Omega_f - L_f \to \mathbb{E} - L$ extends to a holomorphic isomorphism $\phi : \Omega_f \to \Omega$ conjugating f to F.

Another way to say this is that L is the set of tips of a simple elamination Λ, with one leaf for each pair $\ell(q^{\pm})$. And Ω is precisely the Riemann surface obtained from

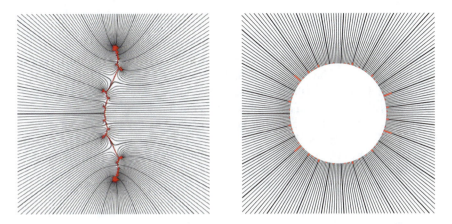

Fig. 4 Vertical leaves in Ω_f and in $\mathbb{E}$ for $f(z) := z^3 + 3z + 3^{-1/2}$

Λ by pinching, together with the 1-form dz/z whose real and imaginary parts are the (derivatives of) height and argument respectively.

When one talks about constructing a Riemann surface by gluing Euclidean polygons, one sometimes says the Riemann surface is built 'from paper' (see e.g. [14]). As a mnemonic therefore, and by abuse of homonymy, we say that Ω is built from *butcher paper*.

In case some critical points are not simple, there might be three (or more) segments in L associated to some (pre)-critical points, and some segment $\ell(q^{\pm})$ associated to a critical point c might be a subsegment of some precritical $\ell(r^{\pm})$ associated to another critical point. Exactly as in the simple case, these sets form the tips of the leaves of an elamination Λ (no longer simple) and $\Omega = \mathbb{E} \mod \Lambda$.

Definition 4.3 (*Dynamical Elamination*) The elamination Λ obtained from f as above is called the *dynamical elamination* associated to f.

If we need to stress the dependence of Λ on f we denote it $\Lambda(f)$.

Lemma 4.4 *The assignment* $\Phi : f \rightarrow \Lambda(f)$ *is a continuous function from* $\mathcal{S}_d$ *to* $\mathcal{EL}$ *that we call the* butcher map.

Proof The Fatou sets Ω_f together with their vertical/horizontal foliations vary continuously as a function of f. Since $\Lambda(f)$ can be recovered from Ω_f under the identification of $\mathbb{E} \mod \Lambda(f)$ with Ω_f, and since we defined the topology on $\mathcal{EL}$ so that the inverse of pinching is continuous, the lemma follows. $\qquad\qquad\square$

5 Formal Shift Space

In this section we shall characterize the dynamical elaminations $\Lambda(f)$ that arise from shift polynomials by the construction in Sect. 4.4, and describe an inverse map. The existence of this inverse is the Realization Theorem 5.4, due essentially to DeMarco–McMullen, although we express things in rather different language.

In this section we use logarithmic coordinates and fix the notation $\log(z) = r + i\theta$ for $z \in \mathbb{E}$, so that $r \in \mathbb{R}^+$ and $\theta \in \mathbb{R}/2\pi\mathbb{Z}$, and we denote the radial segment associated to z by $\ell(r, \theta)$. In (r, θ) coordinates, the map $z \to z^d$ acts as multiplication by d. We call r the *height* and θ the *angle* of the segment $\ell(r, \theta)$.

5.1 Dynamical Elaminations

The geometry and combinatorics of L is best expressed in the language of elaminations. Let's fix the degree d in what follows.

Definition 5.1 (*Critical data*) A (degree d)*critical leaf* is an extended leaf whose tips have angles that are equal mod $2\pi d^{-1}$.

If $C_1, \cdots, C_e$ is a finite set of degree d critical leaves, we say the *critical multiplicity* of C_j is equal to its ordinary multiplicity, minus 1 for every C_k with greater height which shares a pair of ideal points with C_j.

A (degree d) *critical set* is a finite elamination consisting of degree d critical leaves $C_1, \cdots, C_e$ whose critical multiplicities sum to $d - 1$.

The map $z \to z^d$ acts on radial segments by $\ell(r, \theta) \to \ell(dr, d\theta)$. This induces a (partially) defined action on extended leaves, that might reduce multiplicity if distinct tips have angles that differ by a multiple of $2\pi d^{-1}$. If λ is a leaf for which all tips have angles that differ by a multiple of $2\pi d^{-1}$, the image of λ under $z \to z^d$ is undefined. For instance, $z \to z^d$ is undefined on any critical leaf. If P is a leaf, we denote its image under $z \to z^d$ by P^d.

Definition 5.2 (*Dynamical Elamination*) A *dynamical elamination* L is an elamination containing a finite subset of leaves C which is a degree d critical set, and such that $z \to z^d$ maps $L - C$ to L in a d to 1 manner. We say L is *generated by* C.

Figure 5 indicates a simple dynamical elamination of degree 3.

Proposition 5.3 (Dynamical elamination) *Let C be a degree d critical set. Then there is a unique dynamical elamination L generated by C.*

Proof Recall that the notation S^1 mod C denotes the result of pinching the unit circle along C. From the definition of a critical set, S^1 mod C is the union of d disjoint circles, each canonically isomorphic to $\mathbb{R}/\frac{1}{d}\mathbb{Z}$ (with respect to the angle coordinates it inherits from S^1). Thus the map $z \to z^d$ maps each of these circles

Fig. 5 Simple dynamical elamination of degree 3; critical leaves are in red

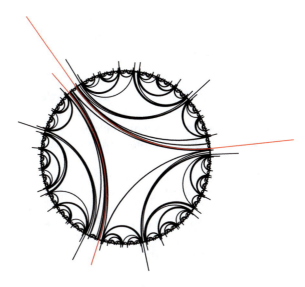

isomorphically to the unit circle. An extended leaf in S^1 mod C canonically pulls back to an extended leaf on the unit circle by taking the preimage of the tips to be the tips of the preimage. We may therefore inductively construct L as the union of L_n where $L_0 = C$ and L_j is obtained from L_{j-1} by taking the preimages of L_j in S^1 mod C and pulling back to an elamination on S^1. Uniqueness is clear. □

We refer to the preimages of the critical leaves as *precritical leaves*, and we say that the *depth* of a precritical leaf P is the number of iterates of the dynamical map which take it to some C_i.

5.2 Realization

Let L be a degree d dynamical elamination generated by C, and let Ω be the Riemann surface obtained from L by pinching. The map $z \to z^d$ induces a degree d proper holomorphic map F from Ω to itself with $d - 1$ critical points counted with multiplicity, which are the endpoints of the tips of the C.

The *Realization Theorem* says that the action of F on Ω is holomorphically conjugate to the action of some (unique) shift polynomial f on its Fatou set.

Theorem 5.4 (Realization) *Let C be a degree d critical set with dynamical elamination L and associated Riemann surface $F : \Omega \to \Omega$. Then there is a unique conjugacy class of degree d polynomial f in the shift locus for which $f|\Omega_f$ is holomorphically conjugate to $F|\Omega$.*

Essentially the same theorem is proved by DeMarco–McMullen [21], Theorem 7.1 although in different language, and with quite a different proof. One distinctive feature of our proof of Theorem 5.4 is that it finds the desired embedding of Ω in $\mathbb{CP}^1$ by a rapidly convergent algorithm; we expect this might be useful e.g. for computer implementation.

Proof The Riemann surface Ω has one isolated puncture (corresponding to ∞) and a Cantor set J of ends (the 'image' of the unit circle under iterated cut-and-paste along $\partial^- L$). The map F extends holomorphically over the isolated puncture; we claim that it also extends (uniquely, holomorphically) over J. The resulting extension will be a degree d holomorphic self-map from a sphere to itself, which is conjugate to a polynomial.

We now explain how to extend the dynamics of F over J holomorphically. Let X be the subset of Ω consisting of points with height $\leq t$ where t is less than the height of any critical leaf, and let Y be the closure of $X - F^{-1}(X)$. Then Y is a (typically disconnected) compact planar surface with outer boundary $\partial^+ Y := \partial X$, and inner boundary $\partial^- Y := \partial Y - \partial^+ Y$. The map $F : \partial^- Y \to \partial^+ Y$ is a d-fold covering map for which every component maps homeomorphically to its image; thus we may define $F_1, \cdots, F_d : \partial^+ Y \to \partial^- Y$ to be branches of F^{-1} with disjoint images whose union is $\partial^- Y$.

Suppose that $\partial^+ Y = \partial X$ has e components. Let D denote the disjoint union of e copies of the unit disk $\mathbb{D}$. We would like to find a holomorphic embedding $\psi : X \to D$, so that $J := D - \psi(X)$ is a Cantor set, and so that F (or, really, its conjugate by ψ) extends holomorphically over J.

Let $\mathcal{T}$ denote the Teichmüller space of holomorphic embeddings $\psi : Y \to D$ taking components of $\partial^+ Y$ to components of ∂D, and normalized to take fixed values on three marked points on each component. We define a *skinning map* $\sigma : \mathcal{T} \to \mathcal{T}$ as follows. Given ψ, we cut out $D - \psi(Y)$ and sew in d copies of D by gluing their boundaries to $\psi(\partial^- Y)$ along the identifications

$$\partial D \xrightarrow{\psi^{-1}} \partial^+ Y \xrightarrow{F_j} \partial^- Y \xrightarrow{\psi} \psi(\partial^- Y)$$

We then uniformize the resulting surface D' to obtain a holomorphic identification $D' \to D$, and the restriction of this uniformization to Y (which we identify with its image in D' under ψ) is $\sigma(\psi)$. The skinning map is holomorphic, and therefore distance non-increasing in the Teichmüller metric. In fact it is evidently strictly distance decreasing; furthermore, orbits are easily seen to be bounded. Thus σ is uniformly strictly distance decreasing, and there is a (unique) fixed point (actually convergence to the fixed point is easy to see directly by considering moduli of accumulating annuli around points of J).

By construction, this fixed point gives the desired embedding of X and extension of F. $\square$

We denote by $\mathcal{DL}_d$ the space of degree d dynamical elaminations, thought of as a subspace of $\mathcal{EL}$. Theorem 5.4 produces a continuous inverse to the butcher map $\Phi : \mathcal{S}_d \to \mathcal{EL}$ called the *realization map* $\Psi : \mathcal{DL}_d \to \mathcal{S}_d$; in particular, the spaces $\mathcal{S}_d$ and $\mathcal{DL}_d$ are homeomorphic.

The location of the tips of the critical leaves define local holomorphic coordinates on $\mathcal{DL}_d$ giving it the structure of a complex manifold. With respect to these coordinates, Φ and Ψ are holomorphic; thus $\mathcal{DL}_d$ and $\mathcal{S}_d$ are isomorphic as complex manifolds.

5.3 Squeezing

There is a free proper $\mathbb{R}$ action on $\mathcal{DL}_d$ which simultaneously multiplies the heights of the critical leaves by some fixed positive real number e^t. We call this transformation *squeezing*, and refer to the $\mathbb{R}$ action as the *squeezing flow*.

Since the squeezing flow is (evidently) proper, it gives $\mathcal{DL}_d$ the structure of a global product:

Corollary 5.5 *Each $\mathcal{DL}_d$ is homeomorphic to a product $\mathcal{DL}_d = X_d \times \mathbb{R}$ where X_d is a real manifold of dimension $2d - 3$.*

For concreteness, we may think of X_d as the subspace of $\mathcal{DL}_d$ where the largest critical height is equal to 1.

5.4 Rotation

If P is a leaf in L, we let $e^{i2\pi t} P$ denote the result of rotating P anticlockwise through t, mod leaves of greater height. This makes sense unless P collides with a leaf of the same height. If P and Q are leaves of different height, the operations of rotating P and rotating Q commute.

If L is a dynamical elamination of degree d with distinct critical leaves, let L_j be the critical leaf C_j and its preimages. Suppose no two critical leaves have heights whose ratio is a power of d; we say L has *generic heights*. Then for a vector $s :=$ $s_1, \cdots s_{d-1}$ of real numbers we can simultaneously rotate all the leaves of each L_j of height h through angle hs_j, mod leaves of greater height; since leaves of the same height are all rotated through the same angle, they never collide and this operation is well-defined. Denote the result by $\mathcal{F}_s L := \cup_j e^{i2\pi h s_j} L_j$.

Lemma 5.6 (Torus orbits) *If L is a degree d dynamical elamination with generic heights $h(C)$, then $\mathcal{F}_s L \in \mathcal{DL}_d$. Furthermore the orbit map $\mathbb{R}^{d-1} \to \mathcal{DL}_d$ factors through a torus $T_L := \mathbb{R}^{d-1}/\Gamma_L$ where Γ_L is contained in $d^{-n}h(C)^{-1}\mathbb{Z}^{d-1}$ for some n.*

Proof By induction, for each precritical leaf P we have $(e^{i\theta}P)^d = e^{i\theta d}P^d$ mod leaves of greater height. Thus $\mathcal{F}_s L$ is a degree d dynamical elamination.

For each critical leaf C_j the angles of C_j vary continuously in a component of S^1 mod leaves of greater height. Since the angles of these leaves of greater height all differ by multiples of d^{-n} for some fixed n, the length of this component is a multiple of $\mathbb{R}/d^{-n}\mathbb{Z}$. The lemma follows. $\square$

6 Degree 2

Our goal in the sequel is to investigate the topology and combinatorics of $\mathcal{S}_d$. As a warm-up, and in order to introduce the main ideas in a relatively clean context, we describe in the next few sections the special cases of degrees 2, 3 and 4. After developing the theory of the past few sections, the case of degree 2 is almost a triviality.

Theorem 6.1 (Douady–Hubbard [18]) *The space $\mathcal{S}_2$ is holomorphically equivalent to a punctured disk.*

Proof A degree 2 dynamical elamination L is generated by a single (necessarily simple) critical leaf C. The tips of C are of the form $\ell(z)$ and $\ell(-z)$ for some $z \in \mathbb{E}$. Since every other leaf of L has smaller height than C, the number z^2 is a continuous function of $\mathcal{DL}_2$, and conversely we can recover C and therefore L from z^2. Hence $\mathcal{DL}_2$ is holomorphically isomorphic to the quotient of $\mathbb{E}$ by ±1. $\square$

Corollary 6.2 *The Mandelbrot Set $\mathcal{M}$ (i.e. the complement of $\mathcal{S}_2$ in $\mathbb{C}$) is connected.*

7 Degree 3

The classic references for the degree 3 case are Branner–Hubbard [7, 8].

7.1 *The Tautological Elamination*

Throughout this section we refer to the *angles* of a leaf P of an elamination as the arguments of the tips divided by 2π; thus angles take values in the circle $S^1 = \mathbb{R}/\mathbb{Z}$.

For some small $\epsilon > 0$ and angles $t, s \in S^1$ let $L(t, s)$ be the degree 3 dynamical elamination with simple critical leaves C_1, C_2 where C_1 has height 1 and angles $\{t, t + 1/3\}$, and C_2 has height $1 - \epsilon$ and angles $\{s, s + 1/3\}$. Note that this forces $s \in (t + 1/3, t + 2/3)$.

If we fix t and vary s in $(t + 1/3, t + 2/3)$, then whenever $3^n s$ is equal to t or $t + 1/3$, the leaf C_2 collides with a leaf P of $L(t, s)$ which is a depth n preimage of C_1. We define an elamination $\Lambda_T(t)$ whose leaves are the union of the leaves P^3 over all P in all $L(t, s)$ of this kind.

Fig. 6 P and P^3 (in blue) have angles $\{5/9, 16/27, 8/9\}$ and $\{2/3, 7/9\}$

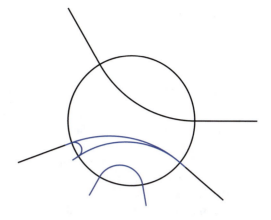

Example 7.1 Let $t = 0$ and $s = 5/9$. Thus C_1 has angles $\{0, 1/3\}$ and C_2 has angles $\{5/9, 8/9\}$. There is a unique leaf P with angles $\{s = 5/9, s'\}$ which collides with C_2 for which $P^9 = C_1$ and neither P nor P^3 crosses C_1 or C_2 (actually, because P is saturated by C_2, it has angles $\{s = 5/9, s', 8/9\}$ but we ignore this point, since the tips with angles $5/9$ and $8/9$ become equal in P^3 and it is the leaf P^3 that is in $\Lambda_T(0)$). The leaf P^3 has angles $\{3s = 2/3, 3s'\}$; since $9s' = 1/3 \mod \mathbb{Z}$, for P^3 not to cross C_1 or C_2 we must have $3s' = 7/9$. Thus, in order for P not to cross C_1 or C_2 we must have $s' = 16/27$. See Fig. 6.

The leaf P^3 with height $1/3$ and angles $\{2/3, 7/9\}$ is therefore a leaf of $\Lambda_T(0)$.

Definition 7.2 (*Tautological Elamination*) Fix $t \in S^1$. The *tautological elamination* $\Lambda_T(t)$ is the union of P^3 over all leaves $P \in L(t, s)$ in the preimage of C_1 over all values of s at which $C_2 \in L(t, s)$ collides with P.

If $P \in L(t, s)$ is a depth n preimage of C_1 that collides with C_2, we refer to its image $P^3 \in \Lambda_T(t)$ as a *depth* $(n - 1)$ *leaf of* $\Lambda_T(t)$.

Proposition 7.3 *For all* t, $\Lambda_T(t)$ *is an elamination. Furthermore,* $\Lambda_T(t + s) = e^{i2\pi hs} \Lambda_T(t)$ *for any* t, s.

Proof As we vary C_2 fixing its height, the preimages of C_1 are occasionally pushed over preimages of C_2 of greater height. But a depth 1 preimage P of C_1 has height $1/3$, which is greater than the height of any preimage of C_2, so P is only pushed over C_2 itself. Since the angles of C_2 differ by $1/3$, pushing P over C_2 does not change its image P^3. So we can simply add P^3 to $\Lambda_T(t)$.

Now imagine shrinking the height of C_2 to $1/3(1 - \epsilon)$ and then varying its angles again. The depth 1 preimages of C_1 pinch the unit circle into smaller circles, and C_2 is confined to a single component. Since C_1 now has height $< 1/3$, the depth 2 preimages Q of C_1 in this component have bigger height than any preimage of C_2, so they stay fixed until they collide with C_2, and we can simply add the Q^3 to $\Lambda_T(t)$. In other words: the depth 2 leaves of $\Lambda_T(t)$ are the cubes of the depth 2 preimages

Fig. 7 Tautological elaminations $\Lambda_T(1/12)$ to depths 1, 2, 3, 4 and 6

of C_1 in the component of S^1 pinched along the depth 1 preimages of C_1 containing C_2. It follows that these leaves are disjoint, and do not cross depth 1 leaves.

Inductively, shrink the height of C_2 to $3^{-n}(1-\epsilon)$. It is confined to a component of S^1 pinched along the depth $\leq n$ preimages of C_1, and as it moves around this component, it collides with some depth $(n+1)$ preimages R of C_1 and we add R^3 to $\Lambda_T(t)$. It follows (as before) that these leaves are disjoint and do not cross leaves of depth $\leq n$. This proves that $\Lambda_T(t)$ is an elamination.

To see how $\Lambda_T(t)$ varies with t, shrink C_2 down to the height of a depth n preimage P it has just collided with. Then rotate C_1 and simultaneously rotate C_2 at speed 3^{-n} (modulo leaves of greater height) so that it continues to collide with P.   □

Figure 7 depicts subsets of the tautological elaminations up to depth six associated to $\theta_1 = 1/12$ in units where the unit circle has length 1.

7.2 Topology of $\mathcal{S}_3$

Let $\Omega(t) = \mathbb{E} \mod \Lambda_T(t)$, and let $D_\infty(t)$ be the subsurface of $\Omega(t)$ of height $\leq 3(1-\epsilon)$. Then $D_\infty(t)$ is a disk minus a Cantor set, and as t varies, the $D_\infty(t)$ vary by 'rotating' the level sets of height h through angle $ht/3$. By Proposition 7.3 this family of motions for $t \in [0, 1]$ induces a mapping class φ of $D(0)$ to itself. The mapping torus N_∞ of φ is the total space of a fiber bundle over S^1 whose fiber over t is $D_\infty(t)$.

Figure 8 shows a tautological elamination $\Lambda_T(5/6)$ and the disk D_∞ obtained by pinching it (to depth 7). These pictures were generated by the program shifty [13] which pinches elaminations recursively one leaf at a time, instead of simultaneously pinching all leaves of fixed depth. Thus the picture of D_∞ is only a combinatorial approximation, and is not conformally accurate.

Theorem 7.4 (Topology of $\mathcal{S}_3$) *The space $\mathcal{S}_3$ is homeomorphic to a product $X_3 \times \mathbb{R}$ where X_3 is the 3-manifold obtained from the 3-sphere S^3 by drilling out a neighborhood of a right-handed trefoil and inserting the mapping torus N_∞, so that the longitude intersects the circle $\partial D_\infty(t)$ at angle t.*

Proof This follows more or less directly from the definitions. Let's examine the subspace Y_3 of X_3 for which $h(C_1) = 1$ and $h(C_2) \leq (1-\epsilon)$. If we fix θ_1 and the height $h := h(C_2)$ then we obtain a (1-dimensional) subspace $\Gamma(\theta_1, h)$ of Y_3. Evidently $\Gamma(\theta_1, h)$ is obtained from the circle of possible θ_2 values $[\theta_1 + 1/3, \theta_1 +$

2/3]/endpoints by suitable cut and paste. By multiplying angles by 3 we can iden-
tify this space of θ_2 values with the unit circle S^1; so $\Gamma(\theta_1, h)$ is obtained from S^1
by cut and paste. We claim it is precisely equal to the result of cut and paste along
the leaves of $\Lambda_T(t)$ of height $> h$.

To see this, think about a component γ of $\Gamma(\theta_1, h)$; its preimage $\tilde{\gamma}$ in S^1 is a union
of segments. The discontinuities of θ_2 in $\Gamma(\theta_1, h)$ occur precisely when C_2 is pushed
over a precritical leaf of C_1 of height $> h$; thus the boundary of each component of
$S^1 - \tilde{\gamma}$ is a precritical leaf P of C_1 so that C_2 collides with P in some dynamical
elamination $L(\theta_1, s)$. But then by definition P^3 is a leaf of $\Lambda_T(\theta_1)$, and all leaves of
$\Lambda_T(\theta_1)$ arise this way. This proves the claim, and shows that Y_3 is homeomorphic to
N_∞.

It remains to show that $X_3 - Y_3$ is homeomorphic to the complement of the right
handed trefoil. For each $h \in (1/3, 1)$ the slice of X_3 for which $h(C_2) = h$ is just a
torus T, with coordinates $\theta_1 \in S^1$ and $\theta_2 \in [\theta_1 + 1/3, \theta_1 + 2/3]$/endpoints. When
$h = 1$ we can no longer distinguish C_1 and C_2, so this torus is quotiented out by
the involution switching θ_1 and θ_2 coordinates; the quotient is a circle bundle over
an interval with orbifold endpoints of orders 2 and 3—see Fig. 9. Thus $X_3 - Y_3$ is a
circle bundle over a disk with two orbifold points, one of order 2 and one of order 3;
this is the standard Seifert fibered structure on $S^3 - \text{trefoil}$. $\qquad\qquad\square$

Fig. 8 Tautological
elamination $\Lambda_T(5/6)$ and the
disk obtained by pinching it

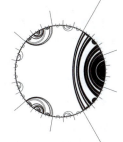
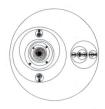

Fig. 9 Quotient of the torus
T by the involution
switching θ_1 and θ_2 is a
circle bundle over an interval
with orbifold endpoints of
orders 2 and 3

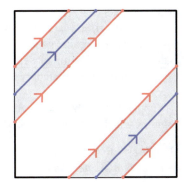

7.3 Geometry and Topology of X_3

Let $\Lambda_T(\theta_1, n)$ denote the finite elamination consisting of the leaves of $\Lambda_T(\theta_1)$ of depth $\leq n$ (i.e. they correspond in the construction of the tautological elamination to depth n preimages of C_1).

Let $\Omega_n(\theta_1)$ be the Riemann surface obtained by pinching $\Lambda_T(\theta_1, n)$ and let $D_n(\theta_1)$ be the subsurface of height $3(1 - \epsilon)$. Then each $D_{n+1}(\theta_1)$ is obtained by pinching $D_n(\theta_1)$ along the depth $(n + 1)$ leaves, and we can think of $D_\infty(\theta_1)$ as the limit. Likewise we can define mapping tori N_n which are $D_n(\theta_1)$ bundles over the θ_1 circle S^1.

Let M_n denote the result of inserting N_n into the right-handed trefoil complement in S^3. Then M_n is a link complement, $S^3 - K_n$ where K_0 is the trefoil itself and each K_{n+1} is obtained from K_n by (a rather simple) satellite of its components. The limit $K_\infty = S^3 - X_3$ is a Cantor set bundle over S^1; one sometimes calls such objects *Solenoids*.

We now state and prove two theorems, which describe S_3 in geometric resp. topological terms. The geometric statement is that S_3 is homotopic to a *locally* CAT(0) *2-complex*. This means a 2-dimensional CW complex (in the usual sense) with a path metric of non-positive curvature; see e.g. [9] for an introduction to the theory of CAT(0) spaces.

The most important corollary of this structure for us is that a locally CAT(0) complex is a $K(\pi, 1)$; the proof is a generalization of the usual proof of the Cartan–Hadamard theorem for complete Riemannian manifolds of nonpositive curvature (which are themselves examples of locally CAT(0) spaces). Thus (for example) $\pi_1(S_3)$ is torsion free, and has vanishing homology with any coefficients in dimension greater than 2.

Theorem 7.5 *(CAT(0) 2-complex)* S_3 *is a* $K(\pi, 1)$ *with the homotopy type of a locally CAT(0) 2-complex.*

Proof Up to homotopy, we can take M_0 to be the *spine* of the trefoil complement; this is the mapping torus of a theta graph by an order three isometry that permutes the edges by a cyclic symmetry. It can be thickened slightly to M_0 by gluing on a metric product (flat) torus times interval. Each M_n has boundary a union of totally geodesic flat tori, and each M_{n+1} is obtained by gluing a flat annulus whose boundary components are parallel geodesics in ∂M_n (circlewise, the endpoints of a leaf of the tautological elamination of depth $(n + 1)$) and then gluing a flat torus times interval on each resulting boundary component to thicken. The union is homeomorphic to X_3.

Simply gluing the spines at each stage without thickening gives a homotopic complex which is evidently CAT(0). $\qquad\square$

Corollary 7.6 $\pi_1(S_3)$ *is torsion-free, and homology with any coefficients vanishes in dimension greater than 2.*

The topological statement is that X_3 is homeomorphic to a Solenoid complement of a particularly simple kind: one obtained as an infinite increasing union of iterated cables.

Theorem 7.7 (Link complement) *The degree* 3 *shift locus* $\mathcal{S}_3$ *is homeomorphic to* $X_3 \times \mathbb{R}$ *where* X_3 *is* S^3 *minus a Solenoid* K_∞ *obtained as a limit of a sequence of links* K_n *where*

(1) K_0 *is the right-handed trefoil; and*
(2) *Each component* α *of* K_n *gives rise to new components* $\alpha_0 \cup \alpha_c$ *of* K_{n+1}, *where* α_0 *is the core of a neighborhood of* α *(i.e. we can think of it just as* α *itself) and* α_c *is a finite collection of* (p_α, q_α) *cables of* α_0, *for suitable* p_α, q_α.

Proof The only thing to prove is the second bullet point. Let α be a component of K_n. The boundary of a tubular neighborhood of α is the mapping torus of a finite collection of boundary circles of $D_n(0)$ which are permuted by the monodromy φ. Let m be the least power of φ that takes one such boundary component $\gamma \subset \partial^- D_n(0)$ to itself. Then φ^m acts on γ by rotation through $2\pi p_\alpha / q_\alpha$.

The depth $(n + 1)$ leaves of Λ_T on the component γ form a finite elamination permuted by φ^m. Think of this as determinining a finite geodesic lamination of $\mathbb{D}$. The complementary components are in bijection with the components γ_j of $\partial^- D_{n+1}(0)$ obtained by pinching γ, and we must understand how φ^m acts on them. A finite order rotation of $\mathbb{D}$ has a unique fixed point—the center. So there is a unique component γ_0 invariant under φ^m, and all the other components are freely permuted with period q_α. Evidently under taking mapping tori γ_0 is associated to the core α_0 and the other γ_j are associated to components α_c which are all (p_α, q_α) cables of α_0. $\square$

Corollary 7.8 (Homology of $\mathcal{S}_3$) H_1 *and* H_2 *of* $\mathcal{S}_3$ *(and of* $\pi_1(\mathcal{S}_3)$*) is free abelian on countably infinitely many generators.* $H_0 = \mathbb{Z}$ *and* $H_n = 0$ *for all* $n > 2$.

In fact, it is possible to get more precise information about the denominators q_α, and in fact we are able to show:

Theorem 7.9 (Powers of 2) *The orbit lengths under* φ *of the cuffs of* D_n *(and hence all denominators* q_α *in Theorem 7.7) are powers of* 2.

In fact, the proof of Theorem 7.9 goes via arithmetic, and will be given in Sect. 9; technically, the proof is a consequence of Theorem 9.20 and Example 9.10. We do not actually know a direct combinatorial proof of this theorem in terms of the combinatorics of the tautological elamination, and believe it would be worthwhile to try to find one. We explore the combinatorics of the tautological elamination further in Sect. 9.5.

The tautological elamination has exactly 3^{n-1} leaves of depth n and therefore $(3^n - 1)/2$ leaves of depth $\leq n$. It follows that D_n is a disk with $(3^n + 1)/2$ holes. However, the monodromy φ permutes these nontrivially, and K_n has one component for each orbit.

The links K_n have 1, 2, 5, 11 components for $n = 0, 1, 2, 3$, though the degrees with which these components wrap around the cores of their parents are quite complicated. Thickened neighborhoods of K_n for $n = 0, 1, 2$ are depicted in Fig. 10.

Fig. 10 Thickened neighborhoods of K_j for $j = 0, 1, 2$. X_3 is homeomorphic to $S^3 - K_\infty$

8 Degree 4 and Above

8.1 Weyl Chamber

As in the case of degree 3, we set $\mathcal{S}_4 = X_4 \times \mathbb{R}$ where X_4 is the quotient of $\mathcal{S}_4$ by the orbits of the squeezing flow.

Order the critical heights with multiplicity so that $h_1 \geq h_2 \geq h_3$ and define a map $\rho : X_4 \to \mathbb{R}^3$ with coordinates $t_j := -\log_4 h_j$. If we identify X_4 with the subspace for which $h_1 = 1$ then $t_1 = 0$ and the image of ρ is the subset of $(t_2, t_3) \in \mathbb{R}^2$ with $0 \leq t_2 \leq t_3$. Another normalization is to set $\sum t_j = 0$ in which case the image of ρ may be identified with the Weyl chamber W associated to the root system A_2.

Within this chamber we have a further stratification. Define $t_{ij} := t_i - t_j$ and refer to the level sets $t_{ij} = n \in \mathbb{Z}$ as *walls*. The walls define a cell decomposition τ of W into right angled triangles with dual cell decomposition τ'.

We shall describe a natural partition of X_4 into manifolds with corners $X_4(v)$, for vertices v of τ, where $X_4(v)$ is defined to be the preimage under ρ of the cell of τ' dual to v. These submanifolds are typically disconnected, and the way their components are glued up in X_4 will give X_4 the structure of a *contractible complex of spaces, modeled on an (incomplete) $\tilde{A}_2$ building.*

8.2 Two Partitions

Let's suppose critical leaves are simple, and we label them C_j compatibly with the ordering on heights.

There are two combinatorially distinct ways for C_1 to sit in the circle: the angles of the segments are either antipodal, or they are distance $1/4$ apart (remember we are working in units where the circle has total length 1). When $h(C_1)$ is strictly larger than the other $h(C_j)$ the leaf C_1 is the unique leaf of greatest height. Thus the difference of the angles is locally constant; it follows that the subset of X_4 where $h(C_2) < 1$ is disconnected. In fact, it is easy to see it has exactly two components according to the placement of C_1.

Where C_1 is an antipodal leaf, it pinches the unit circle into two circles of length $1/2$, each bisected by one of C_2 and C_3. The restriction of the dynamical elamination in each of each of these length $1/2$ circles is symmetric under the antipodal map.

When C_1 is not antipodal, it pinches the unit circle into circles of length $1/4$ and $3/4$, with C_2 and C_3 both contained in the longer circle. The leaf C_2 pinches this circle into circles of length $1/2$ and $1/4$, and C_3 divides the length $1/2$ circle antipodally.

8.3 Monkey Prisms, Monkey Turnovers

Let's fix a generic (t_2, t_3) in the interior of W, so that none of $t_2, t_3, t_3 - t_2$ are integers. Denote the fiber of ρ over (t_2, t_3) by $T(t_2, t_3)$. These fibers are disjoint union of 3-tori, orbits of the $\mathbb{R}^3$ action $\mathcal{F}_s$ on $\mathcal{DL}_4$ described in Lemma 5.6. These tori piece together to form a product throughout each open triangle of τ. We let θ_j (taking values in $\mathbb{R}^3$ mod a suitable lattice) denote angle coordinates on one of these tori.

As we pass through a wall where some $t_{ij} \in \mathbb{N}$, circle factors in these tori pinch as follows. The angle coordinates θ and the log height coordinates t determine a dynamical elamination. When $t_{ij} = n$ the circle parameterized by θ_i is pinched along the precritical leaves of C_j of depth n. As we move around in the fiber, the dynamical elamination varies by a rotation, so the way in which the θ_i circle pinches depends only on which component we are in, and the value of the local coordinates θ_j with $j < i$. In other words, the structure locally is that of a certain kind of iterated fiber bundle called a *monkey bundle*.

Recall from Definition 3.6 the terms monkey pants and monkey Morse functions.

Definition 8.1 (*Monkey bundle*) A *monkey bundle* of *order n* consists of the following data:

(1) A finite sequence of fiber bundles $\Omega_2 \to E_2 \to S^1$ and $\Omega_j \to E_j \to E_{j-1}$ for $3 \le j \le n$ where each Ω_j is a monkey pants;
(2) a map $\pi_j : E_j \to [0, 1]$ whose restriction to each Ω_j fiber is monkey Morse; and such that
(3) if $E := E_n$ is the total space, and $\pi : E \to [0, 1]^{n-1}$ denotes the map whose factors restrict to π_j on each E_j, then for each j the image of the critical points in the Ω_j fibers is a collection of affine hyperplanes.

The cube $[0, 1]^{n-1}$ together with the hyperplanes which are the images of fiberwise critical points under π should be thought of as a *graphic* in the sense of Cerf theory; see e.g. [15]. We say that a curve in $[0, 1]^{n-1}$ crosses a hyperplane of the graphic *positively* if it corresponds to the positive direction in the factor $\pi_j : E_j \to [0, 1]$ to which the hyperplane is associated.

Definition 8.2 (*Monkey prism; monkey turnover*) Suppose E is a monkey bundle with projection $\pi : E \to [0, 1]^{n-1}$. Suppose $\Delta \subset [0, 1]^{n-1}$ is a convex polyhedron for which there is a vertex $v \in \Delta$ so that the ray from v to every other point in Δ crosses the graphic in the positive direction. Then we call $P := \pi^{-1}(\Delta)$ a *monkey prism*.

Suppose $\pi : P \to \Delta$ is a monkey prism, and some collection of finite groups act on some boundary strata of P preserving π. Then the quotient space Q of P together with the data of its induced projection to Δ is called a *monkey turnover*.

Lemma 8.3 (Prism is $K(\pi, 1)$) *A monkey prism of order n is a $K(\pi, 1)$ with the homotopy type of an n-complex. A monkey turnover of order n has the homotopy type of an n-complex.*

Proof A monkey pants is homotopic to a graph, and iterated fibrations of $K(\pi, 1)$s are $K(\pi, 1)$s. Thus a monkey bundle is a $K(\pi, 1)$ with the homotopy type of an n-complex.

The universal cover $\tilde{E}$ of a monkey bundle E is a (noncompact) manifold with corners, and interior homeomorphic to a product $\mathbb{R}^2 \times \cdots \times \mathbb{R}^2 \times \mathbb{R}$ where each $\mathbb{R}^2$ factor has a singular foliation with leaf space an oriented tree.

If $F \subset E$ is a monkey prism associated to a polyhedron $\Delta \subset [0, 1]^{n-1}$ then the preimage $\tilde{F} \subset \tilde{E}$ is bounded in each $\mathbb{R}^2$ factor by a collection of lines of the foliation, and is homeomorphic to a disjoint union of $\mathbb{R}^2$s. As we move along a straight ray in Δ from the distinguished vertex we might cross hyperplanes of the graphic, but by hypothesis we only cross in the positive direction. As we cross a hyperplane, the part of $\tilde{F}$ in some $\mathbb{R}^2$ fibers splits apart, but pieces can never recombine; thus $\tilde{F}$ is homeomorphic to $\mathbb{R}^{2n-1}$ so that F is also a $K(\pi, 1)$ with the homotopy type of an n-complex.

Since orbifolding is compatible with π, a monkey turnover also has the homotopy type of an n-complex. $\square$

From the description of the fibers of ρ and how they pinch as we cross a wall, the following is immediate:

Lemma 8.4 *Let Δ be a cell of the dual cellulation τ'. Then $\rho^{-1}(\Delta)$ is a disjoint union of monkey prisms and monkey turnovers with respect to the map ρ.*

Figure 9 is a simple example of the way a fiber can be quotiented in a monkey turnover.

There does not seem to be any obvious reason why monkey turnovers in generality should be $K(\pi, 1)$s. However it will turn out that the turnovers that occur in the partition of X_4 are $K(\pi, 1)$s. The reason for this is subtle, and only proved in Sect. 9.

There is another natural cellulation κ of W associated to the subset of walls of the form $t_{i1} \in \mathbb{N}$; i.e. the walls of the integer lattice in $\mathbb{R}^2$. They decompose W into squares and right-angled triangles. Let κ' be the dual cellulation; the cells of κ' are triangles, squares and rectangles, and the cells of κ' are in bijection with the cells of τ'. Since τ and κ have the same set of vertices, there is a bijection between the top dimensional cells of τ' and $'\kappa'$.

In the sequel it will be convenient to compare the monkey prisms and turnovers associated to τ' with those associated to κ'.

Lemma 8.5 (Equivalent Cells) *Let K and T be cells of the cellulations κ' and τ' associated to a vertex v. Then the components of $\rho^{-1}(K)$ and of $\rho^{-1}(T)$ are homeomorphic, and are isotopic inside X_4.*

Proof There is an isotopy of the frontiers of the cells from one to the other which never introduces any new tangency with the graphic. Since fibers are arranged in a product structure away from the graphic, the lemma follows.   □

The prisms and turnovers associated to cells of κ' are naturally homeomorphic to the *moduli spaces* introduced in Sect. 9.3.

8.4 $K(\pi, 1)$

Decompose W into cells dual to the cellulation by walls; note that typical cells (those dual to interior vertices of W) are hexagons. The preimage under ρ of each of these cells is a disjoint union of monkey prisms and monkey turnovers, and the walls in each cell are the graphic. Thus X_4 is a *complex of spaces* in the sense of Corson [16]. The associated complex is built from copies of cells of τ according to the pattern of inclusion of connected components; thus it is locally modeled on an $\tilde{A}_2$ *building*, which comes with an immersion to W. See e.g. Brown [10] for an introduction to the theory of buildings.

Theorem 8.6 (Complex of spaces) *X_4 is a complex of monkey prisms and monkey turnovers over a contractible complex B.*

Proof The direction of pinching is transverse to the walls, so there is a unique path in the building from every point to the origin projecting to a ray in W.   □

In retrospect, the inductive picture of X_3 we obtained in Sect. 7 as an infinite union of knot and link complements, exhibits it as a complex of monkey prisms and monkey turnovers (actually, only one monkey turnover) over a contractible $\tilde{A}_1$ building (i.e. a tree).

The next theorem is the analog in degree 3 of Theorem 7.5.

Theorem 8.7 ($K(\pi, 1)$) *S_4 is a $K(\pi, 1)$ with the homotopy type of a 3-complex.*

We have already seen that the monkey prisms (and consequently also monkey turnovers) in X_4 have the homotopy type of 3-complexes. The same is therefore true of X_4.

X_4 is assembled from monkey prisms and monkey turnovers associated to the vertices of B. The edges and triangles are associated to lower dimensional monkey prisms and turnovers included as facets in the boundary. The monkey prisms and their boundary strata are all $K(\pi, 1)$s by Lemma 8.3, and the inclusions of boundary strata are evidently injective at the level of π_1. It remains to show that the same holds for the monkey turnovers.

We defer the proof of this to Sect. 9, but for the moment we give some examples to underline how complicated the monkey turnovers can be.

Fig. 11 One of the two monkey turnovers associated to the vertex $(1, 1)$ is a Y_2 bundle over S^1, where Y_2 is built from five pieces associated to the configurations indicated in the figure. The first two pieces are $K(B_3, 1)$s and the last three are $K(\mathbb{Z}^2, 1)$s. C_1 and its preimages with greater height than C_2, C_3 are in red

Example 8.8 ($K(B_4, 1)$) The turnover associated to the vertex $(0, 0)$ homotopy retracts onto the fiber $\rho^{-1}(0, 0)$. This is the (3 real dimensional) configuration space of degree 4 dynamical elaminations with all critical leaves of height 1. This turns out to be a spine for the configuration space of 4 distinct unordered points in $\mathbb{C}$; i.e. it is a $K(B_4, 1)$ (an analogous statement holds in every degree). There are several ways to see this; one elegant method is due to Thurston, and explained in [29]. We shall see a quite different and completely transparent demonstration of this fact in Sect. 9.

Example 8.9 (*Star of David*) There are two monkey turnovers associated to the vertex $(1, 1)$ of τ in W, corresponding to the two combinatorially distinct ways for C_1 to sit in S^1.

When C_1 is antipodal, the leaves C_2 and C_3 sit on either side and do not interact with each other. For each fixed value of C_1 the other two leaves vary as a product $P \times P$ of pairs of pants. Monodromy around the C_1 circle switches the two factors by an involution.

When C_1 is not antipodal, the leaves C_2 and C_3 may interact, and the topology is significantly more complicated. This component is also a bundle over S^1 whose fiber is a certain 4-manifold Y_2 that we call the *Star of David* (the explanation for the name will come in Sect. 9). It is built from five pieces; two of these pieces are homotopic to trefoil complements (i.e. they are $K(B_3, 1)$s). The other three pieces are homotopic to tori, which attach to the other components along a subspace homotopic to a wedge of two circles; in other words this decomposition does *not* form an injective complex of $K(\pi, 1)$s. In fact, the fundamental group of Y_2 is obtained from the free product of two B_3s by adding three commutation relations. The five pieces are illustrated in Fig. 11.

8.5 Degree d

Most of what we have done in this section generalizes to degree d readily. Set $\mathcal{S}_d = X_d \times \mathbb{R}$, and order critical heights with multiplicity so that $1 = h_1 \geq h_2 \geq \cdots \geq h_{d-1}$. Define $\rho : X_d \to \mathbb{R}^{d-2}$ with coordinates $t_j := -\log_d h_j$ for $j = 2, \cdots, d-1$. The image of X_d is the Weyl chamber W, which is partitioned by walls $t_{ij} \in \mathbb{Z}$ where $t_{ij} := t_i - t_j$ into the cells of a cell decomposition τ with dual decomposition τ'. If we identify $\mathbb{R}^{d-2}$ affinely with the subspace of $\mathbb{R}^{d-1}$ with coordinates summing to 0, then τ becomes the *symplectic honeycomb*; see e.g. Coxeter [17]. For example, in degree 5 the cells of τ are regular tetrahedra and octahedra, and the cells of τ' are regular rhombic dodecahedra.

Let κ be the cellulation defined only by the subset of walls t_{i1} and let κ' be the dual cellulation. Then we have:

Lemma 8.10 (Equivalent Cells) *Let K and T be cells of the cellulations κ' and τ' associated to a vertex v. Then the components of $\rho^{-1}(K)$ and of $\rho^{-1}(T)$ are homeomorphic, and are isotopic inside X_d.*

Theorem 8.11 (Complex of spaces) *S_d is a complex of monkey prisms and monkey turnovers over a contractible complex locally modeled on an $\tilde{A}_{d-2}$-building.*

Theorem 8.12 (Homotopy dimension) *S_d has the homotopy type of a $(d-1)$-complex (i.e. a complex of half the real dimension of S_d as a manifold).*

The proofs are all perfectly analogous to the proofs of Lemma 8.5, Theorem 8.6 and (the relevant part of) Theorem 8.7.

8.6 Tautological Elaminations

It is straightforward to generalize Definition 7.2 to higher degree for the critical leaves of least height. Fix $C_1, C_2, \cdots, C_{d-2}$ at heights $h_1 \geq h_2 \cdots h_{d-2}$, and let C_{d-1} at height $h_{d-2} - \epsilon$ vary. Every time C_{d-1} collides with a leaf P which is a preimage of C_j for $j < d-1$ we add P^d to the tautological elamination.

It is harder to decide on a definition for the other critical leaves. This is because the elamination associated to C_j depends on the fixed locations of C_k with $k < j$ and an *equivalence class* of fixed locations of C_k with $k > j$. We explain.

Definition 8.13 (*Degree d Tautological Elaminations*) Fix a degree d and an index $1 < i \leq d-1$. Fix locations of leaves C_j for $j \neq i$ where the C_j with $j < i$ have heights $h_1 \geq h_2 \geq \cdots h_{i+1}$, and the C_j with $j > i$ have height 0. We shall define the leaves of the tautological elamination $\Lambda_T(C)$ associated to $C := C_1, \cdots \hat{C}_i \cdots C_{d-1}$ of depth n. Insert C_i somewhere at height $h_{i+1} - \epsilon$ compatibly with the other leaves, and construct the leaves of the dynamical elamination associated to the critical data $C \cup C_i$ which are preimages of C_j up to depth n. As we vary C_i, the leaves C_j with $j < i$ stay fixed but the C_j with $j > i$ are pushed over C_i and over preimages of higher depth critical leaves. Whenever C_i collides with a preimage P of a higher C_j we add P^d to the tautological elamination.

The C_j with $j > i$ are 'hidden parameters'; we need them to determine the location of the preimages of greater height, but they do not themselves contribute any leaves to Λ_T.

As the angles of C_j, $j < i$ vary by a vector of parameters t (and C_j, $j > i$ are pushed over by this motion) the tautological elaminations vary by the flow $\mathcal{F}_t$.

Within each monkey prism the pinching is described by these tautological elaminations. Let's fix a cell τ' dual to a vertex v where $t_j = n_j$ and a monkey prism which is a component of $\rho^{-1}(\tau')$. The way in which the fiber Ω_i over $C_{<i} \in E_{i-1}$ pinches depends on which component we are in; implicitly, this choice of component determines an equivalence class of the location of C_j with $j > i$ and therefore determines a tautological elamination. The depth $\leq n_i - n_j$ preimages of the C_j in the tautological elamination describe the pinching of Ω_i as a function of $C_{<i}$. The proof is perfectly parallel to that of Theorem 7.4.

8.7 Completed Tautological Elamination

Fix $C := C_1, \cdots, \hat{C}_i, \cdots, C_{d-1}$ as above. It is possible to define a suitable 'completion' of the tautological elamination $\Lambda_T(C)$ as follows.

Definition 8.14 (*Completed Tautological Elamination*) Fix d and C as above. In the construction of the tautological elamination, set the formal height of C_i to be equal to 0, and define $\mathcal{L}_n$ to be the set of leaves of the form P^d where P is a depth n preimage of C_i that collides with C_i itself.

Although they have height 0, the $\mathcal{L}_n$ have a well-defined vein in $\mathbb{D}$. Note that some pairs of leaves of $\mathcal{L}_n$ cross each other in $\mathbb{D}$. Nevertheless we can think of $\mathcal{L}_n$ as a closed subset of the space of geodesic leaves in $\mathbb{D}$ and take the lim sup $\mathcal{L}_\infty :=$ lim sup$_{n \to \infty} \mathcal{L}_n$ (i.e. there is a leaf in $\mathcal{L}_\infty$ for each convergent sequence of leaves in a subsequence of the $\mathcal{L}_n$). Then we define the *completed* tautological elamination associated to C to be $\bar{\Lambda}_T(C) := \Lambda_T(C) \cup \mathcal{L}_\infty$.

The leaves of $\bar{\Lambda}_T(C) - \Lambda_T(C)$ are called *flat* since they have height 0, to distinguish them from the *ordinary* leaves of $\Lambda_T(C)$.

Theorem 8.15 (Limit is lamination) *The vein of $\bar{\Lambda}_T(C)$ is a geodesic lamination (i.e. leaves of $\mathcal{L}_\infty$ do not cross $\Lambda_T(C)$ or each other).*

The proof of this will appear in a forthcoming paper.

Pinching along $\bar{\Lambda}_T(C)$ is the same as pinching along $\Lambda_T(C)$, since the flat leaves all have height zero, so do not actually intrude into $\mathbb{E}$. However, it *does* make sense to pinch the closure $\bar{\mathbb{E}} \subset \mathbb{C} \cup \infty$ along $\bar{\Lambda}_T(C)$, exactly as before by cut and paste along the tips of $\Lambda_T(C)$, and then by quotienting the endpoints of the flat leaves to single points. Let's call the result $\bar{\Omega}_T(C)$. Because we added limits in the definition of $\bar{\Lambda}_T(C)$, $\bar{\Omega}_T(C)$ is Hausdorff. It is a compactification of $\Omega_T(C)$ away from ∞, by locally connected spaces (isolated points or monotone quotients of circles).

Notice that this construction is non-vacuous even when $d = 2$; it reproduces Thurston's quadratic geolamination [28], which is a proposed topological model for the boundary of the Mandelbrot set (proposed, since it is famously unknown if the Mandelbrot set is locally connected).

Thus it seems reasonable to conjecture that the boundary components of $\overline{\Omega}_T(C)$ should parameterize (modulo the question of local connectivity) the boundaries of the components of the complement of $\mathcal{S}_d$ in the slice associated to C. Compare with [2].

9 Sausages

In this section we introduce a completely new way to see the pieces in the building decomposition of X_d via algebraic geometry. It will turn out that the monkey prisms and monkey turnovers in X_d all become homeomorphic (after taking a product with an interval) to (rather explicit) complex affine varieties—moduli spaces of certain objects called *sausage shifts*.

9.1 Sausages: The Basic Idea

Everyone likes sausages. Now we will see them made. The basic idea is illustrated in Fig. 12.

A dynamical elamination is a machine that, by a process of repeatedly pinching leaves in order of height, extrudes a long, complicated Riemann surface Ω (a Fatou set); by tying this Riemann surface off at periodic values of $-\log_d h$, we decompose it into manageable genus zero chunks: sausages.

Thus the Riemann surface Ω is tied off into a tree of sausages, and the dynamics of F on Ω decomposes into polynomial maps between the sausages, whose moduli spaces are described by (elementary) algebraic geometry.

Fig. 12 Making sausages

9.2 Definitions

9.2.1 Tagged Points

Let f be a holomorphic map between open subsets of $\mathbb{C}$ taking p to q. If $f'(p)$ is nonzero, df is a $\mathbb{C}$-linear isomorphism from T_p to T_q. Thus after scaling by a suitable positive real number, it induces an isometry of unit tangent circles. We denote these unit tangent circles by U and the induced map as $Uf : U_p \to U_q$.

If p is a critical point of multiplicity m, then f maps infinitesimal round circles centered at p to infinitesimal round circles centered at q by a degree $(m + 1)$ covering. By abuse of notation we write $Uf : U_p \to U_q$ for this map. In holomorphic coordinates for which f is $z \to z^{m+1}$ this map is just multiplication by $(m + 1)$ on U_0 (really we are using an implicit identification between the tangent space T_q and its $(m + 1)$st tensor power).

Definition 9.1 (*Tagged Point*) A *tagged point* is a point p together with an element $u_p \in U_p$. The *zero tag* is the point $0 \in \mathbb{C}$ together with the unit vector $u_0 \in U_0$ tangent to the positive real axis.

If f is a holomorphic map taking a tagged point p to a tagged point q we say it *preserves tags* if $Uf(u_p) = u_q$. If f is a holomorphic function, a *tagged root* is a tagged point p with $f(p) = 0$ for which $Uf(u_p)$ is the zero tag.

9.2.2 Sausages

Let T be a locally finite rooted tree. Every vertex v but the root has a unique *parent*— the unique vertex adjacent to v on the unique embedded path in T from v to the root. If w is the parent of v we say v is a child of w. Every edge of T is *oriented* from child to parent.

Definition 9.2 (*Bunch of sausages*) Let T be a locally finite rooted tree. A *bunch of sausages* over T is an infinite nodal genus 0 Riemann surface S made from a copy of $\mathbb{CP}^1$ for each vertex v of T (the *sausages*, which we denote $\mathbb{CP}^1_v$) and for each v a finite set of *marked tagged points* $Z_v \subset \mathbb{CP}^1_v - \infty$ and a bijection σ from the children of v to the set Z_v, so that if w is a child of v, the point ∞ in the sausage $\mathbb{CP}^1_w$ is attached to the point $\sigma(w) \in Z_v \in \mathbb{CP}^1_v$.

If T is a rooted tree, for each vertex w of T there is a rooted subtree $T_w \subset T$ with root w. If S is a bunch of sausages over T, then $S_w \subset S$ denotes the bunch of sausages associated to the subtree T_w.

A *morphism* between rooted trees T, T' is a simplicial map $\tau : T \to T'$ taking roots to roots, and directed edges to directed edges. Thus if w is a child of v, the image $\tau(w)$ is a child of $\tau(v)$.

Definition 9.3 (*Augmentation*) If T is a rooted tree, the *augmentation* of T, denoted T', is the rooted tree obtained from T by adding a new root v' and an edge from the root v of T to v'. If S is a bunch of sausages over T, the *augmentation* of S, denoted S', is the bunch of sausages over T' obtained by attaching $\mathbb{CP}^1_{v'}$ along $0 = Z_{v'}$ to ∞ in Z_v.

Definition 9.4 (*Polynomial*) Let S be a bunch of sausages over a locally finite tree T. A *degree d polynomial p* is a degree d tagged holomorphic map from S to its augmentation S' over a morphism $\tau : T \to T'$. This means that for every vertex w of T there is a polynomial map $p_w : \mathbb{CP}^1_w \to \mathbb{CP}^1_{\tau(w)}$ of degree d_w in normal form taking Z_w to $Z_{\tau(w)}$, and so that

(1) if v is the root, the polynomial p_v has degree d and its roots are exactly $Z_v \subset \mathbb{CP}^1_v$, and furthermore as tagged points Z_v are tagged roots of p_v;
(2) the root polynomial p_v has more than one root; i.e. p_v is not the polynomial z^d;
(3) for every vertex w with $\tau(w) = u$ the map $p_w : \mathbb{CP}^1_w \to \mathbb{CP}^1_u$ takes Z_w to Z_u as tagged points, and Z_w is the entire preimage $p_w^{-1}(Z_u)$; and
(4) if w is the child of u with $\sigma(w) = z \in Z_u \subset \mathbb{CP}^1_u$ then the degree d_w of the polynomial p_w is equal to the multiplicity of z as a preimage under p_u.

The second bullet point is a kind of nondegeneracy condition: if the root polynomial p_v were z^d, then S would already be the augmentation of some other sausage polynomial.

Lemma 9.5 *Let S be a bunch of sausages over T, and let $p : S \to S'$ be a degree d polynomial over a morphism $\tau : T \to T'$. Then for every vertex $w' \in S'$ the sum of degrees $\sum_{\tau(w)=w'} d_w = d$, and every point in S' has exactly d preimages, counted with multiplicity.*

Proof This is true for the root vertex by bullet (1) from Definition 9.4, and by induction by bullets (3) and (4). □

This lemma justifies the terminology 'polynomial map'.

Definition 9.6 Let S be a bunch of sausages over T, and p a polynomial map of degree d. Let w be a vertex of T, and let $c \in \mathbb{CP}^1_w - \infty$ be a critical point for p_w. We say c is a *genuine* critical point if c is not in Z_w and is *false* otherwise.

We say p is a *degree d shift polynomial* and (S, p) is a *degree d sausage shift* if there are exactly $d - 1$ genuine critical points, counted with multiplicity.

Bullet (2) in the Definition 9.4 is equivalent to saying that the root sausage contains at least one genuine critical point.

If p is a shift polynomial, there is a minimal finite rooted subtree $U \subset T$ containing all the genuine critical points. Thus for $w \in T - U$, every polynomial p_w is degree 1; since it is in normal form it is the identity map $p_w(z) = z$.

Corollary 9.7 *Let S be a bunch of sausages over T, and let p be a degree d shift polynomial. Then the space $\mathcal{E}(T)$ of ends of T is a Cantor set, and the action of p on $\mathcal{E}(T)$ is conjugate to the one-sided shift on right-infinite words in a d-letter alphabet.*

9.2.3 Isomorphism of Polynomials

The definition of a sausage polynomial includes data in the form of tags that is essential if we want to construct a map from sausage polynomials to shift polynomials, as we shall do in Sect. 9.4. In order for this map to be injective we must quotient out by a (finite) equivalence relation that we now explain.

Let S be a bunch of sausages over a tree T, and let p be a degree d polynomial as in Definition 9.4. Let u be a vertex of T, let $z \in Z_u \subset \mathbb{CP}^1_u$, and let w be the child of u with $\sigma(w) = z$. If z is a critical point of p_u of multiplicity m then p_w has degree $m + 1$; i.e. the degree of p_u near z agrees with the degree of p_w near infinity. In the sequel we will 'cut open' $\mathbb{CP}^1_u$ at z and $\mathbb{CP}^1_w$ at infinity, and sew together the two resulting boundary circles in a dynamically compatible way, lining up the tag at z in $\mathbb{CP}^1_u$ with the positive real axis at infinity in $\mathbb{CP}^1_w$.

The tag at z maps under p_u to the tag at $p_u(z)$; thus given p_u and the choice of tag at $p_u(z)$ we have freedom in the choice of a compatible tag at z: different choices differ by multiplication by an $(m + 1)$st root of unity ζ. If we multiply the tag at z by ζ, we must at the same time change the coordinates on $\mathbb{CP}^1_w$ by multiplication by ζ. Changing coordinates on $\mathbb{CP}^1_w$ inductively affects the data associated to w and the subtree T_w and its preimages under p in the obvious way. For example, $p_w(z)$ is replaced by $p_w(\zeta^{-1}z)$, the marked points Z_w are replaced by their preimages ζZ_w, etc.

We say two sausage shifts are *isomorphic* if they are related by a finite sequence of modifications of this sort. There are $\prod_{w \in T} \prod_{z \in Z_w} (m(z) + 1)$ polynomials in an isomorphism class, where $m(z)$ is the multiplicity of z as a critical point of p_w, and where the product is taken over all $z \in T_w$ for all $w \in T$. Note that for a sausage shift, this product is finite, since all but finitely many p_w have degree 1.

9.3   Moduli Spaces

For each fixed combinatorial type of degree d sausage shift, there is an associated *moduli space* of isomorphism classes with the given combinatorics, parameterized locally by the coefficients of the vertex polynomials p_w of degrees > 1. We shall see in Theorem 9.15 that moduli spaces for sausage shifts with generic heights have complex dimension $d - 1$, and in fact they have the natural structure of iterated fiber bundles of complex affine varieties.

This is best explained by examples.

Example 9.8 (*Degree 2*) The root polynomial p_v is of the form $z^2 - c$ for some nonzero c. Since every other polynomial has degree 1 (and is therefore the identity function z) S is a rooted dyadic tree, where each parent has two children attached at $\pm\sqrt{c}$. The moduli space of such sausages is evidently $\mathbb{C}^*$. This is homeomorphic (but *not* holomorphically isomorphic) to S_2.

Example 9.9 (*Distinct roots*) The simplest case in every degree d is that the root polynomial p_v has distinct roots. Then every other polynomial has degree 1 and S is a rooted d-adic tree, where each parent has d children attached at the roots of p_v. Thus the moduli space is a discriminant complement, and hence a $K(B_d, 1)$.

Example 9.10 (*Degree 3*) Suppose the root polynomial p_v has two roots, so it is of the form $p_v := (z - c)^2(z + 2c) = z^3 - 3c^2z + 2c^3$ with c nonzero. The root vertex v has two children u, w where u is attached at the double root c (say). Then $p_w = z$ and p_u has degree 2. Either 0 is a genuine critical point for p_u, or p_u is of the form $z^2 + c$ or $z^2 - 2c$. In the latter case u has two children u', w' where u' is attached at 0 and this chain of critical roots $u, u', u^{(2)}, u^{(3)}, \cdots$ continues until $p_{u^{(n)}} := z^2 + x$ has a genuine critical point (or equivalently, $x \in \mathbb{C} - Z_t$ where p takes the vertex $u^{(n)}$ to t). The moduli space is a bundle over $\mathbb{C}^*$ (parameterized by the choice of c) whose fiber is $\mathbb{C} - Z_t$.

Notice that the points of Z_t are obtained from $c, -2c$ by repeatedly pulling back under double branch covers of the form $z \to z^2 + c_j$ where c_j is one of the preimages pulled back so far. The monodromy acts on each of these double branch covers either trivially or by permuting some of the preimages in pairs. It follows that every orbit of the monodromy on Z_t has length a power of 2.

Example 9.11 (*Star of David*) Suppose that the root polynomial in degree 4 has one simple root and one triple root; i.e. the root polynomial is $p_v := (z - c)^3(z + 3c)$ with c nonzero. The root has two children u, w where u is attached at the triple root c (say). The simplest case is when c and $-3c$ are regular values for p_u. Then the moduli space is a bundle over $\mathbb{C}^*$ whose fiber is Y_2, the space of degree 3 polynomials $z^3 + pz + q$ for which two specific distinct complex numbers (in this case c and $-3c$) are regular values. It turns out that this moduli space is homotopic to the monkey turnover described in Example 8.9.

The general structure of moduli spaces should now be starting to become clear. To make a precise statement, we introduce the notion of a *Hurwitz Variety*:

Definition 9.12 (*Hurwitz Variety*) A *degree d Hurwitz variety* is an affine complex variety of the following form. Fix a finite set $Q \subset \mathbb{C}$ and a conjugacy class of representation σ from $\pi_1(\mathbb{CP}^1 - Q)$ to the symmetric group S_d.

The *Hurwitz Variety* $H(Q, \sigma, d)$ is the space of degree d normalized polynomials of the form $f(z) := z^d + a_2 z^{d-2} + \cdots + a_d$ for which $f : \mathbb{C} \to \mathbb{C}$ is a degree d branched cover whose monodromy around q is conjugate to $\sigma(q)$ for all $q \in Q$.

For a permutation σ let $|\sigma| = d -$ number of orbits. Thus $|\sigma(q)|$ is the multiplicity of q as a critical value of f, for each $q \in Q$ and each $f \in H(Q, \sigma, d)$. We establish some basic properties of these varieties:

Proposition 9.13 (Basic Properties) *Hurwitz varieties $H(Q, \sigma, d)$ satisfy the following basic properties:*

(1) the dimension of $H(Q, \sigma, d)$ is equal to $d - 1 - \sum_q |\sigma(q)|$;

(2) $H(Q, \sigma, d)$ is connected if its dimension is positive;
(3) if there is a homeomorphism from $\mathbb{CP}^1$ to $\mathbb{CP}^1$ taking Q to Q' and conjugating σ to σ' then $H(Q, \sigma, d)$ is homeomorphic to $H(Q', \sigma', d)$.

Proof The first bullet (i.e. dimension count) is elementary.

If we choose a finite subset $P \subset \mathbb{C} - Q$ and extend σ to P then we can build a degree d branched cover of $\mathbb{CP}^1$ over $P \cup Q$ with monodromy σ at $P \cup Q$. The genus of this branched cover depends only on σ. Thus the family of covers which are connected and genus 0 form a bundle over the space of pairs $Q \cup P, \sigma$ of a particular combinatorial type, and it is an exercise in finite group theory to show that these fibers are connected when they have positive dimension. Each $H(Q, \sigma, d)$ is a finite branched cover of the associated fiber (the Riemann surface determines the polynomial up to finite ambiguity); this proves the second bullet.

To prove the third bullet, let's modify our homeomorphism $\varphi : \mathbb{CP}^1 \to \mathbb{CP}^1$ by an isotopy so that it is equal to the identity in a neighborhood of ∞, and is K-quasiconformal for some K. For each $f \in H(Q, \sigma, d)$ we can pull back the Beltrami differential $\mu := \bar{\partial}\varphi / \partial\varphi$ to $f^*\mu$ and let $\phi : \mathbb{CP}^1 \to \mathbb{CP}^1$ uniquely solve the Beltrami equation for $f^*\mu$, normalized to be tangent to the identity at infinity to second order. Then $\psi(f) := \varphi f \phi^{-1}$ is a normalized polynomial, and by construction it is in $H(Q', \sigma', d)$. Letting f range over $H(Q, \sigma, d)$ defines a homeomorphism $\psi : H(Q, \sigma, d) \to H(Q', \sigma', d)$ as desired. $\square$

Example 9.14 (*Discriminant Variety*) If we set $Q = \{0\}$ and σ the map to the identity element, then $H(\{0\}, \mathrm{id}, d)$ is the space of degree d polynomials in normal form with simple roots. In other words, $H(\{0\}, 0, d)$ is the complement of the discriminant variety, and is a $K(B_d, 1)$.

Theorem 9.15 (Moduli spaces) *Every moduli space of a degree d sausage shift of a fixed combinatorial type is an algebraic variety over $\mathbb{C}$ which has the structure of an iterated bundle whose base and fibers are all Hurwitz varieties. Furthermore, it has dimension $d - 1$.*

Proof Consider a vertex w with parent u and image $v = \tau(w)$. There is a polynomial $p_w : \mathbb{CP}^1_w \to \mathbb{CP}^1_v$ whose degree is equal to the multiplicity of u as a preimage under p_u. The points Z_w are the preimages of Z_v under p_w, and the number and multiplicity of these points depends on the monodromy of p_u as a branched cover around Z_v. Thus for a fixed combinatorial type, the polynomials p_w vary in a Hurwitz Variety whose data is determined by the polynomials in vertices above w. Changing a tag changes the coordinates on the Hurwitz variety by a (finite) automorphism. Thus the moduli space is an iterated bundle as claimed. $\square$

9.3.1 $K(\pi, 1)$s

Hurwitz varieties can apparently be quite complicated, topologically. But at least in low degree we have the following theorem, which is by no means obvious, and which I personally find rather startling:

Theorem 9.16 (*CAT*(0) 2-complex) *Every connected Hurwitz variety* $H(Q, \sigma, 3)$ *is a* $K(\pi, 1)$ *with the homotopy type of a locally CAT*(0) 2-complex.

Proof If any point in Q is a critical value the dimension is 1 or 0 and H is either homotopic to a graph or to a finite set of points. So the only interesting case is when Q is a finite set and σ is the constant map to the identity permutation. In other words, if $|Q| = n$, then $H(Q, \mathrm{id}, 3)$ is the (two complex dimensional) space Y_n of degree 3 polynomials $z^3 + pz + q$ for which the points in Q are regular values. We show these have the homotopy type of locally CAT(0) 2-complexes (and are therefore $K(\pi, 1)$s).

First we describe the topology. By the third bullet of Proposition 9.13 we can take Q to be the set of nth roots of unity. Then $Y_n = \mathbb{C}^2 - V$, where V is the hyperplane in $\mathbb{C}^2$ with coordinates p, q for which $\prod_j (-4p^3 - 27(q - \zeta^j)^2) = 0$. By a linear change of coordinates, we can replace this hyperplane by $\prod_j (x^3 - (y - \zeta^j)^2) = 0$.

V intersects the plane $x = 0$ in exactly the nth roots of unity. We foliate the complement of this plane by (real 3-dimensional) open solid tori $S^1 \times \mathbb{C}$ thought of as a bundle over the circle $|x| = t$, and let V_ϵ denote the intersection with V. If we cutoff $|y|$ at some big T, then we get another solid torus $|y| = T$, $|x| \leq t$ and the union is an S^3. When $|x| = \epsilon$ is small and positive, V_ϵ splits into a union of n trefoils T^j_ϵ (in this S^3), each obtained as a narrow cable of the circle $y = \zeta^j$. The part of Y_n in the domain $|x| \leq \epsilon$ is homotopic to a wedge of n copies of a $K(B_3, 1)$, one for each trefoil.

When $2|x|^{3/2} = |\zeta^j - \zeta^k|$ the trefoils T^j and T^k intersect at three points, and when $|x|$ increases past this value, they become linked. There are no other intersections. The link of a crossing (in $\mathbb{C}^2$) is a Hopf link, and the result of pushing across each such crossing attaches a space to Y_n, homotopic to a 2-torus, attached along a subspace homotopic to a wedge of two circles. In other words, it attaches a 2-cell, whose boundary kills the relator which is the commutator of two meridian circles linking the trefoils at the point of intersection.

For each pair of trefoils T^j, T^k, we may choose Garside generators for $\pi_1(S^3 - T^j)$ corresponding to these meridian circles (the Garside presentation for B_3 is of the form $\langle a, b, c \mid ab = bc = ca \rangle$). Thus each pair of trefoils contributes a subgroup of $\pi_1(Y_n)$ of the form

$$\langle a, b, c, x, y, z \mid ab = bc = ca, xy = yz = zx, [a, x] = [b, y] = [c, z] = 1 \rangle$$

However if we follow this chain of relations around a sequence of three trefoils T^j, T^k, T^l for which j, k, l are positively oriented in $\mathbb{Z}$ mod n (say), the intersection points of each pair of trefoils is successively displaced by a rotation so that the holonomy of this chain of displacements rotates one third of the way around. Thus for a triple of trefoils with Garside generators (a, b, c), (n, m, o) and (x, y, z), the commutation relations take the form

$$[a, n], [b, m], [c, o], [n, x], [m, y], [o, z], [x, b], [y, c], [z, a]$$

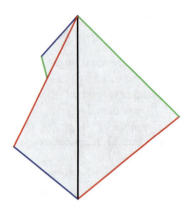

Here is another way of packaging the same information. Build a graph with vertices at the $3n$th roots of unity, and with edges straight line segments between each pair of roots whose ratio is a 3rd root of unity. Then $\pi_1(Y_n)$ is generated by the edges of this graph, with relations that each triple of edges that form a (n equilateral) triangle are Garside generators for a B_3, and each pair of disjoint edges commutes. Furthermore, Y_n is homotopic to the presentation 2-complex associated to this presentation. We shall show this 2-complex (or: a closely related and homotopic complex) can be given a CAT(0) structure.

Actually, there is a beautiful trick, that I learned from Jon McCammond, arising from his work with Tom Brady [5] on the construction of CAT(0) *orthoscheme complexes* for (certain) braid groups. First replace each Garside presentation $\langle a, b, c \mid ab = bc = ca \rangle$ by a presentation of the form $\langle a, b, c, d \mid ab = bc = ca = d \rangle$. A presentation complex can be built from three triangles with edges abd^{-1} etc. The trick is to make these *right angled* regular Euclidean triangles—i.e. to set the lengths of a, b, c to be 1, and the length of d to be $\sqrt{2}$. Let K denote the resulting complex (see Fig. 13), and let K' be the complex built from n copies of K (one for each B_3) and one Euclidean square with edge length 1 for each commutation relation as above. We claim the resulting complex is CAT(0).

Let's see why. The complex K (and K' for that matter) has one vertex; since these complexes are 2-dimensional and Euclidean, we just need to check that the link of the vertex has no loop of length $< 2\pi$. The link L of the vertex of K is a *theta graph*, with three edges of length π. The intersections with the long edge d are the vertices of the theta graph, and the intersections with the edges a, b, c give rise to six points (let's call these *short points*), each at distance $\pi/4$ from some vertex.

The link L' of K' is obtained from n disjoint copies of L by gluing a 4-cycle with edges of length $\pi/2$ for each commutation relation. Each such 4-cycle can be thought of as a complete bipartite graph on two sets of two points, and each pair of points is attached to distinct short points in a copy of L. Since short points in L are all distance π apart, no cycle in the graph associated to two B_3s and their commutators has length $< 2\pi$. By the way, this shows that $\pi_1(Y_2)$ is CAT(0).

There is a simplicial map from L' to the complete graph K_n with edges all of length $\pi/2$ which just collapses each copy of L to a point, and identifies edges between the same pair of copies of L. A loop γ in L' of length $< 2\pi$ would project to a (possibly immersed) simplicial 'loop' in K_n of simplicial length at most 3. If the projection has simplicial length 0 then γ is contained in a copy of L which we already know has no loops of length $< 2\pi$. Simplicial length 1 is impossible. If the projection of γ has simplicial length 2 in K_n then γ is contained in a subgraph formed from a pair of copies of L which (as we have just discussed) has no loops of length $< 2\pi$. If the projection of γ has simplicial length 3 then it passes through a cycle of three Ls, and because of the holonomy described above, a length $3\pi/2$ path in γ has endpoints on the same copy of L but at *different* short points. Thus γ has length at least 2π and we are done. $\qquad\qquad\qquad\qquad\qquad\qquad\qquad\qquad\qquad\qquad\qquad\qquad\qquad\qquad\qquad$ □

Together with Theorem 9.15 this immediately implies:

Corollary 9.17 *Every moduli space in degree 4 is a $K(\pi, 1)$.*

Theorem 9.18 *Is every Hurwitz Variety a $K(\pi, 1)$? Is every Hurwitz Variety homotopic to a CAT(0) complex?*

9.4 The Sausage Map

Let $\hat{\mathcal{S}}_d$ be the subspace of $\mathcal{S}_d$ for which $\log_d(h_1) \in (-1/2, 1/2)$, where h_1 is the greatest critical height, and $\log_d$ denotes log to the base d. This space is homeomorphic to $X_d \times (-1/2, 1/2)$, which is to say it is homeomorphic to $\mathcal{S}_d$ itself.

For $f \in \hat{\mathcal{S}}_d$ let $L \in \mathcal{DL}_d$ be the dynamical elamination associated to f by the butcher map, and let Ω be the Riemann surface obtained by pinching L (so that Ω is canonically isomorphic to the Fatou set of f).

Let $\hat{\Omega}$ be the subspace of Ω with $\log_d(h) \leq 1/2$ and let S be the quotient space of $\hat{\Omega}$ obtained by collapsing each component with $\log_d(h) \in 1/2 + \mathbb{Z}$ to a point (which we call a *node*).

Each component V of S minus its nodes can be given a (branched) Euclidean structure with horizontal coordinate θ and vertical coordinate $\nu(h)$, where $\nu : \mathbb{R}^+ - d^{1/2+\mathbb{Z}} \to \mathbb{R}$ is a function that stretches each interval $(d^{n-1/2}, d^{n+1/2})$ to $\mathbb{R}$ by a homeomorphism (depending on n) in such a way that the map $z \to z^d$ on Ω is conformal in the new coordinates.

Let's explain this in terms of $\mathbb{E}$. In logarithmic coordinates h, θ we can think of $\mathbb{E}$ as a half-open Euclidean cylinder which is the product of the unit circle with the positive real numbers. The map $z \to z^d$ becomes multiplication by d, which we denote $\times d$. For each integer n let I_n denote the open interval $(d^{n-1/2}, d^{n+1/2})$ and let A_n be the annulus in $\mathbb{E}$ where $h \in I_n$, and let $A := \cup_n A_n \subset \mathbb{E}$. Thus $\mathbb{E} - A$ is a countable set of circles with $\log_d(h) \in 1/2 + \mathbb{Z}$. Thus $\times d$ takes A_n to A_{n+1} for each n.

Choose (arbitrarily) an orientation-preserving diffeomorphism $\nu_0 : I_0 \to \mathbb{R}$ and for each n define $\nu_n : I_n \to \mathbb{R}$ by $\nu_n(h) := d^n \nu_0(d^{-n}h)$. Thus, by induction, $\nu_{n+1}(dh) = d\nu_n(h)$ for all n and all $h \in I_n$. Then define $\mu : A \to S^1 \times \mathbb{R}$ by $\mu(\theta, h) = (\theta, \nu_n(h))$ for $(\theta, h) \in A_n$. Thus μ semi-conjugates $\times d$ on A to $\times d$ on $S^1 \times \mathbb{R}$. If we identify $S^1 \times \mathbb{R}$ conformally with $\mathbb{C}^*$ by exponentiating, then μ semi-conjugates $\times d$ on A to $z \to z^d$ on $\mathbb{C}^*$. If we keep a separate 'copy' $\mathbb{C}_n^* := \mu(A_n)$ for each n, then we could say that μ conjugates $\times d$ on A to the self-map of $\cup_n \mathbb{C}_n^*$ that sends each $\mathbb{C}_n^*$ to $\mathbb{C}_{n+1}^*$ by $z \to z^d$.

The components of S minus its nodes are obtained from the A_n by cut and paste along segments of L, an operation which respects the Euclidean structure both in h, θ and $\nu(h), \theta$ coordinates

With respect to this branched Euclidean structure, the closure of each V (i.e. putting the nodes back in) is a compact Riemann surface; in fact, it is isomorphic to $\mathbb{CP}^1$, and it is natural to choose ∞ to be the (unique) node of greatest height. Thus S becomes an infinite nodal genus 0 Riemann surface. Furthermore although the quotient map from $\hat{\Omega}$ to S is very far from being holomorphic, the map $z \to z^d$ on Ω does descends to a *holomorphic* map p from S to its augmentation giving S the structure of a bunch of sausages, and p the structure of a degree d shift polynomial. Notice that the images of the critical points are precisely the genuine critical points of the sausage polynomial.

Tags are defined at the nodes by identifying the unit tangent bundle at each node with a circle in Ω, and inductively pulling back tags compatibly with the dynamics of $z \to z^d$ so that the tag at the unique node in the root of the augmentation corresponds to the argument $\theta = 0$ (this is well-defined, since θ takes values in $\mathbb{R}/\mathbb{Z}$ in the subspace of Ω with h greater than any critical height).

Theorem 9.19 (Sausage map) *The sausage map is surjective, and is 1–1 on the subspace of $\hat{S}_d$ for which no critical leaf C_j has $\log_d(h_j) \in 1/2 + \mathbb{Z}$. This subspace maps bijectively to the set of isomorphism classes of degree d sausage shifts.*

Proof It suffices to define a (continuous) inverse. Here is the construction. Cut open a bunch of sausages along its set of nodes and sew in a copy of the unit tangent circle U at each point. Reparameterize the vertical coordinate on each component by the inverse of μ (here we must choose the correct branch depending on the combinatorial distance to the root). Each component becomes in this way a bordered Riemann surface. The point ∞ in each $\mathbb{CP}^1_w$ gets a canonical tag, namely the vector associated to the positive real axis. Thus we obtain a collection of bordered surfaces, so that each border is a round circle with a tag, and we glue these up respecting arguments and tags. By the definition of isomorphism, the gluing is well-defined on an isomorphism class of sausage shift. The result is a complete planar Riemann surface Ω with one punctured end, and the sausage polynomial descends to a degree d self-map on Ω with $(d-1)$ critical points, counted with multiplicity. By the Realization Theorem 5.4 this is the Fatou set of a unique shift polynomial. $\square$

Theorem 9.20 (Monkey pieces are moduli spaces) *The sausage map induces homeomorphisms from $(-1/2, 1/2)$ times the open monkey prisms and monkey turnovers*

arising in the decomposition in Theorem 8.11 to the moduli spaces of degree d sausage shifts of each fixed combinatorial type.

Proof The factor of $(-1/2, 1/2)$ comes from the difference between $\hat{S}_d$ and X_d, via orbits of the squeezing flow.

This is a consequence of Theorem 9.19 and Lemma 8.10. Explicitly: the components of the images of the sausage map are (up to this factor of $(-1/2, 1/2)$) both the subspaces of $\rho^{-1}(W)$ in the preimage of the cells κ', and at the same time they are (by definition) the moduli spaces of generic degree d sausage shifts. □

Together with Corollary 9.17 and the discussion in Sect. 8.4 this completes the proof of Theorem 8.7. Moreover, together with Example 9.10, this completes the proof of Theorem 7.9.

9.5 Sausages and Combinatorics of the Tautological Elamination

We have already seen (Example 9.10) that moduli spaces reveal nontrivial information about the tautological elamination. Let Λ_T denote the (depth 3) tautological elamination for some fixed θ_1, and let $\Lambda_{T,n}$ denote the subset of leaves of depth $\leq n$. We have seen that monodromy permutes the components of S^1 mod $\Lambda_{T,n}$ in such a way that the orbits have length a power of 2.

We claim that these components all have *lengths* of the form $2^m/3^n$ for various m. Fix a sausage polynomial as in Example 9.10 where the root vertex v has $Z_v = c, -2c$, and where there is a chain of vertices $u_1, \cdots, u_n$ mapping to vertices $v = t_1, t_2, \cdots, t_n$ by polynomials $p_j := z^2 + c_j$ so that 0 is a fake critical point for each $j < n$ (i.e. $c_j \in Z_{t_j}$) and a genuine one for $j = n$ (c_n is not in Z_{t_n}).

The components of S^1 mod $\Lambda_{T,n}$ associated to sausages of this combinatorial form are in bijection with the points of Z_{t_n}. Each $w \in Z_{t_n}$ maps by a succession of polynomials of degrees 1 or 2 until it reaches c or $-2c$ (which themselves are mapped to 0 by p_v). The length of a component is multiplied by $1/3$ when we pull back a regular value, and is multiplied by $2/3$ when we pull back a critical value. This proves the claim.

Table 1 shows the number of components of length $\ell/3^n$ at each depth n (omitted entries are zeroes).

Note that there is a unique component with $\ell = 2^n$ for each n; this corresponds to the sausages for which $t_j = u_{j-1}$, $p_{u_1} = z^2 + c$ and $p_{u_j} = z^2$ for $1 < j < n$. The next biggest components have length $2^{\lfloor n/2 \rfloor}/3^n$.

The (n, ℓ) entry in this table is the number of components of length $\ell/3^n$ at depth n. If we denote this entry $N(n, \ell)$ then

$$\sum_\ell N(n, \ell) = (3^n + 1)/2 \text{ and } \sum_\ell N(n, \ell) \cdot \ell = 3^n$$

Table 1 Number of components of length $\ell/3^n$ at depth n

$n\backslash\ell$	1	2	2^2	2^3	2^4	2^5	2^6	2^7	2^8	2^9	2^{10}	2^{11}	2^{12}
0	1												
1	1	1											
2	3	1	1										
3	7	6	0	1									
4	21	16	3	0	1								
5	57	51	13	0	0	1							
6	171	149	39	5	0	0	1						
7	499	454	117	23	0	0	0	1					
8	1497	1348	360	66	9	0	0	0	1				
9	4449	4083	1061	207	41	0	0	0	0	1			
10	13347	12191	3252	591	126	17	0	0	0	0	1		
11	39927	36658	9738	1799	370	81	0	0	0	0	0	1	
12	119781	109898	29292	5351	1125	240	33	0	0	0	0	0	1

Example 9.21 (*Recurrence*) Eric Rains observed the recurrence relation in the first column that

$$N(2n, 1) = 3 \cdot N(2n - 1, 1) \text{ and } N(2n + 1, 1) = 3 \cdot N(2n, 1) - 2 \cdot N(n, 1)$$

(a similar recurrence holds in higher degree). The proof of this is surprisingly delicate, and will appear in a forthcoming paper [12].

Example 9.22 (*Short ℓ sequences*) One reason to be interested in the lengths of components of S^1 mod $\Lambda_{T,n}$ is that it gives us insight into the geometry of the *complement* of $\mathcal{S}_3$. Actually, it is easy enough to describe the picture in arbitrary degree.

For each degree d the *shift complement* is $\mathbb{C}^{d-1} - \mathcal{S}_d$. When critical points are simple, order them by height $h_1 \geq h_2 \geq \cdots h_{d-1}$, and define a *butcher's slice* $B(C_1, \cdots, C_{d-2})$ to be the subset of $\mathcal{S}_d$ with $C_1, \cdots, C_{d-2}$ fixed and $h_{d-1} < h_{d-2}$. There is a tautological elamination $\Lambda_T(C_1, \cdots, C_{d-2})$ (see Sect. 8.6), and the result of pinching gives a Riemann surface Ω_T for which the subset of height $< h_{d-2}$ is holomorphically equivalent to B.

For the sake of simplicity, let's suppose $1 = h_1 = h_2 = \cdots h_{d-2}$ so that the leaves of Λ_T of depth n all have height d^{-n}. A chain of successive components of S^1 mod $\Lambda_{T,n}$ with lengths $\ell_n \cdot d^{-n}$ determines a system of disjoint annuli in the butcher's slice with moduli $1/\ell_n$. So if $\sum_n 1/\ell_n$ diverges (for instance, if the sequence ℓ_n is bounded), the modulus goes to infinity and the end of B converges to an *isolated* point in the complement of the shift locus. Call such an end of B a *small end*. All but countably many of the (uncountable) ends of B are small.

As we exit a small end of B, points in the Julia set collide in the limit to give rise to a non-shift Cantor Julia set (c.f. Example 2.6; also compare with Branner

[6]). The local path component of the shift complement containing this limit point has complex dimension $d - 2$, and is parameterized by the escaping critical points. There are uncountably many of these local path components, parameterized locally by the small ends of B.

Dragging critical points off to the (Cantor) Julia set one by one defines a nested sequence of holomorphic submanifolds of the shift complement, each parameterized by the remaining escaping critical points. When $C_{j+1} \cdots C_{d-2}$ have been dragged off to J_f, we can define a butcher's slice by fixing $C_1, \cdots , C_{j-1}$ and letting C_j vary; this slice is the subset of height $< h_{j-1}$ in the Riemann surface $\Omega_T(C)$ associated to the tautological elamination $\Lambda_T(C)$ with critical data $C := C_1, \cdots , \hat{C}_j, \cdots C_{d-1}$ for a suitable equivalence class of $C_{j+1}, \cdots , C_{d-2}$ (see Sect. 8.6). Small ends of these butcher's slices locally parameterize the space of these $(j - 1)$-dimensional submanifolds.

10   Fundamental Groups

10.1   Braid Groups

Let $\Delta_d \subset \mathbb{C}^{d-1}$ be the discriminant variety, parameterizing degree d polynomials in normal form $z^d + a_2 z^{d-2} + \cdots a_d$ with multiple roots. The group $\pi_1(\mathbb{C}^{d-1} - \Delta_d)$ acts as permutations of these roots; the permutation representation is a surjective map from $\pi_1(\mathbb{C}^{d-1} - \Delta_d)$ to the symmetric group S_d.

This map is very far from being injective. A loop in $\mathbb{C}^{d-1} - \Delta_d$ defines not just a permutation of roots, but a *braid*: the mapping class represented by the combinatorial manner in which the points move around each other. In other words, there is a *monodromy representation* Mon : $\pi_1(\mathbb{C}^d - \Delta_d) \to B_d$ where B_d is Artin's *braid group* on d strands. Forgetting the braiding determines a surjection Art : $B_d \to S_d$.

Thus we obtain a factorization

$$\pi_1(\mathbb{C}^{d-1} - \Delta_d) \xrightarrow{\text{Mon}} B_d \xrightarrow{\text{Art}} S_d$$

where the first map is an isomorphism, and the second indicates that B_d is functorially obtained from S_d by the algebraic process of *Artinization*.

10.2   Shift Automorphisms

Let Σ_d denote the space of right-infinite words on a d letter alphabet; i.e. $\Sigma_d := \{1, \cdots , d\}^{\mathbb{N}}$. This is a Cantor set in the product topology, and the shift σ acts as a d to 1 expanding map. Let $\hat{S}_d$ denote the group $\hat{S}_d := \text{Aut}(\Sigma_d, \sigma)$; i.e. the group of homeomorphisms of the Cantor set commuting with the shift. See e.g. [4] for background.

In [1], Blanchard–Devaney–Keen showed that the natural map $\pi_1(\mathcal{S}_d) \to \hat{S}_d$ is surjective, in every degree d. As before, this is very far from being injective (as we shall shortly see).

Monodromy defines a representation Mon : $\pi_1(\mathcal{S}_d) \to \mathrm{Mod}(\mathbb{C} - \text{Cantor set})$, but this map is certainly not an isomorphism, since $\pi_1(\mathcal{S}_d)$ is countable whereas $\mathrm{Mod}(\mathbb{C} - \text{Cantor set})$ has the cardinality of the continuum. Actually, the image can be lifted to $\mathrm{Mod}(\mathrm{Disk} - \text{Cantor set})$, since all shift polynomials (in normal form) are tangent to second order near infinity. Let's denote the image by $\hat{B}_d$.

Forgetting the braiding defines a surjective homomorphism $A : \mathrm{Mod}(\mathrm{Disk} - \text{Cantor set}) \to \mathrm{Aut}(\text{Cantor set})$, and the image of $\hat{B}_d$ is $\hat{S}_d$. I proved (see [11]) that $\mathrm{Mod}(\mathrm{Disk} - \text{Cantor set})$ is left-orderable, and therefore torsion-free, whereas $\hat{S}_d$ is generated by torsion.

In any case we have a factorization of the Blanchard–Devaney–Keen map as

$$\pi_1(\mathcal{S}_d) \xrightarrow{\text{Mon}} \hat{B}_d \xrightarrow{A} \hat{S}_d$$

Neither map seems easy to understand. On the other hand, with Juliette Bavard and Yan Mary He we were able to show:

Theorem 10.1 (Bavard–Calegari–He) *In degree 3 the map Mon* : $\pi_1(\mathcal{S}_3) \to \hat{B}_3$ *is an isomorphism.*

The proof of this theorem shall (hopefully!) appear in a forthcoming paper. The most optimistic conjecture I can make is:

Conjecture 10.2 (*Monodromy Conjecture*) The map Mod : $\pi_1(\mathcal{S}_d) \to \hat{B}_d$ is an isomorphism in every degree.

The only real evidence I have in favor of this conjecture is that it is not obviously falsified by the simplest cases I was able to fully analyze.

If $Y = H(Q, \sigma, e)$ is a Hurwitz variety, the preimage of Q under $f \in Y$ is a finite subset of $\mathbb{C}$ whose cardinality is constant as a function of f, and therefore we obtain a monodromy map $M : \pi_1(Y) \to B_n$ for suitable n depending on Y. If Y is a Hurwitz variety that arises as a fiber of a moduli space, the image of $\pi_1(Y) \to \pi_1(\mathcal{S}_d) \to \hat{B}_3$ factors through this B_n, so the monodromy conjecture implies that the maps M are injective. In fact, at least in low dimensions, the monodromy conjecture is *equivalent* to injectivity on these pieces, since both $\pi_1(\mathcal{S}_d)$ and $\hat{B}_d$ are built up in understandable ways from these pieces (this is how Theorem 10.1 is proved).

In any case, this is something we can test, since the groups $\pi_1(Y)$ and B_n are rather explicit, especially in low degree.

Example 10.3 (*Star of David*) The 'hard' pieces in degree 4 are the Star of David and its generalizations as discussed in Theorem 9.16.

Recall the moduli space Y_2 from Example 9.11, and the description of its fundamental group in Theorem 9.16. This fundamental group (let's call it G) has a presentation

$$G := \langle a, b, c, x, y, z \mid ab = bc = ca, xy = yz = zx, [a, x] = [b, y] = [c, z] = 1 \rangle$$

The monodromy map to B_6 arises by thinking of the generators as the edges of a Star of David in the plane, and taking each generator to the braid that cycles the endpoints of the edge around each other in a narrow ellipse contained in a neighborhood of the edge.

There is an isometric embedding from the CAT(0) complex for G described in Theorem 9.16 to the Brady–McCammond complex for B_6, which has been shown to be CAT(0) by Haettel–Kielak–Schwer [24]. If the image is totally geodesic, this would imply that $G \to B_6$ is injective. This seems quite plausible, but we have not checked it.

Acknowledgements I would like to thank Laurent Bartholdi, Juliette Bavard, Pierre Deligne, Laura DeMarco, Yan Mary He, Sarah Koch, Jeff Lagarias, Chris Leininger, Jon McCammond, Curt McMullen, Madhav Nori, Kevin Pilgrim, Eric Rains, Alden Walker, Henry Wilton and the anonymous referee for their help. Most of what I know about polynomial dynamics (which is not much) I learned from Sarah and from Curt at various points in time.

I would also like to extend thanks to the students who attended the graduate topics course I taught on this material at the University of Chicago in Winter 2019, and to Sam Kim who solicited some talks and a paper for the celebration of the 25th anniversary of the founding of KIAS, and without whom I might have never been sufficiently motivated to write any of this up.

References

1. P. Blanchard, R.L. Devaney, L. Keen, The dynamics of complex polynomials and automorphisms of the shift. Invent. Math. **104**, 545–580 (1991)
2. A. Blokh, L. Oversteegen, R. Ptacek, V. Timorin, Laminational models for some spaces of polynomials of any degree. Mem. Amer. Math. Soc. **265**(1288) (2020)
3. L. Böttcher, The principal laws of convergence of iterates and their application to analysis (Russian). Izv. Kazan. Fiz.-Mat. Obshch. **14**, 137–152 (1904)
4. M. Boyle, J. Franks, B. Kitchens, Automorphisms of the one-sided shift and subshifts of finite type. Erg. Thy. Dyn. Sys. **10**, 421–449 (1990)
5. T. Brady, J. McCammond, Braids, posets and orthoschemes. Algebr. Geom. Topol. **10**(4), 2277–2314 (2010)
6. B. Branner, Cubic polynomials: turning around the connectedness locus, in Topological Methods in Modern Mathematics, ed. by L. Goldberg, A. Phillips (Publish or Perish 1993), pp. 391–427
7. B. Branner, J. Hubbard, The iteration of cubic polynomials. I. The global topology of parameter space. Acta Math. **160**(3–4), 143–206 (1988)
8. B. Branner, J. Hubbard, The iteration of cubic polynomials. II. Patterns and parapatterns. Acta Math. **169**(3–4), 229–325 (1992)
9. M. Bridson, A. Haefliger, *Metric Spaces of Non-positive Curvature*. Grund. der Math. Wiss., vol. 319 (Springer, Berlin, 1999)
10. K. Brown, *Buildings* (Springer, Berlin, 1988)

11. D. Calegari, Circular groups, planar groups, and the Euler class, in *Proceedings of the Casson Fest*. Geometry & Topology Monographs, vol. 7 (Geometry & Topology Publication, Conventry, 2004), pp. 431–491
12. D. Calegari, *Combinatorics of the Tautological Lamination*. To appear
13. D. Calegari, Shifty. Computer program; Source available on request
14. A. de Carvalho, T. Hall, Riemann surfaces out of paper. Proc. Lond. Math. Soc. (3) **108**(3), 541–574 (2014)
15. J. Cerf, La stratification naturelle des espaces de fonctions différentiables réelles et le théorème de la pseudo-isotopie. Inst. Haut. Études Sci. Publ. Math. **39**, 51–73 (1970)
16. J. Corson, Complexes of groups. Proc. Lond. Math. Soc. (3) **65**(1), 199–224 (1992)
17. H. Coxeter, *Regular Polytopes*, 3rd edn. (Dover Publications Inc, New York, 1973)
18. A. Douady, J. Hubbard, Itération des polynômes quadratiques complexes. C. R. Acad. Sci. Paris Sér. I Math. **294**(3), 123–126 (1982)
19. L. DeMarco, Dynamics of rational maps: a current on the bifurcation locus. Math. Res. Lett. **8**(1–2), 57–66 (2001)
20. L. DeMarco, Combinatorics and topology of the shift locus, in *Conformal Dynamics and Hyperbolic Geometry*. Contemporary Mathematics, vol. 573 (American Mathematical Society, Providence, 2012), pp. 35–48
21. L. DeMarco, C. McMullen, Trees and the dynamics of polynomials. Ann. Sci. Éc. Norm. Supér. (4) **41**(3), 337–382 (2008)
22. L. DeMarco, K. Pilgrim, The classification of polynomial basins of infinity. Ann. Sci. Éc. Norm. Supér. (4) **50**(4), 799–877 (2017)
23. L. Goldberg, L. Keen, The mapping class group of a generic quadratic rational map and automorphisms of the 2-shift. Invent. Math. **101**(2), 335–372 (1990)
24. T. Haettel, D. Kielak, P. Schwer, The 6-strand braid group is CAT(0). Geom. Dedicata **182**, 263–286 (2016)
25. C. McMullen, Braiding of the attractor and the failure of iterative algorithms. Invent. Math. **91**(2), 259–272 (1988)
26. J. Milnor, *Dynamics in One Complex Variable*, 3rd edn. Annals of Mathematics Studies, vol. 160 (Princeton University Press, Princeton, 2006)
27. W. Thurston, *Thurston's Notes*, MSRI http://library.msri.org/books/gt3m/
28. W. Thurston, On the geometry and dynamics of iterated rational maps, in *Complex Dynamics*, ed. by D. Schleicher, N. Selinger and with an appendix by Schleicher (A. K. Peters, Wellesley, 2009), pp. 3–137
29. W. Thurston, H. Baik, Y. Gao, J. Hubbard, T. Lei, K. Lindsey, D. Thurston, Degree *d* invariant laminations. *What's next? The mathematical legacy of William P. Thurston*. Annals of Mathematics Studies, vol. 205 (Princeton University Press, Princeton, 2020), pp. 259–325
30. N. Vlamis, Big mapping class groups: an overview, in *The Tradition of Thurston* (Springer, Cham, 2020), pp. 459–496

Printed in the United States
by Baker & Taylor Publisher Services